Advances in Industrial Control

Springer
Berlin
Heidelberg
New York
Barcelona
Budapest
Hong Kong
London
Milan
Paris
Tokyo

R.A. Hyde

H_∞ Aerospace Control Design

A VSTOL Flight Application

With 139 Figures

 Springer

Dr Richard Alden Hyde, MA, PhD
Cambridge Control Limited, Newton House, Cambridge Business Park,
Cowley Road, Cambridge CB4 4WZ, UK

Cover Illustration: Ch. 3 Fig.1. GVAM in transition to wing-borne flight

ISBN 978-1-4471-3051-2

British Library Cataloguing in Publication Data
Hyde, Richard Alden
   H∞ Aerospace Control Design: VSTOL Flight Application. – (Advances in Industrial
   Control Series) I. Title II. Series
   629.1352
ISBN 978-1-4471-3051-2   ISBN 978-1-4471-3049-9 (eBook)
DOI 10.1007/978-1-4471-3049-9

Library of Congress Cataloging in Publication Data
A catalog record for this book is available from the Library of Congress

© Springer-Verlag London Limited 1995
Softcover reprint of the hardcover 1st edition 1995

Typesetting: Camera ready by author

69/3830-543210 Printed on acid-free paper

To my parents, my brother Jon, Karen and Liz

SERIES EDITORS' FOREWORD

The series *Advances in Industrial Control* aims to report and encourage technology transfer in control engineering. The rapid development of control technology impacts all areas of the control discipline. New theory, new controllers, actuators, sensors, new industrial processes, computer methods, new applications, new philosophies,..., new challenges. Much of this development work resides in industrial reports, feasibility study papers and the reports of advanced collaborative projects. The series offers an opportunity for researchers to present an extended exposition of such new work in all aspects of industrial control for wider and rapid dissemination.

The present text considers the application of one of the most successful of the new generation multivariable control design techniques. The H_∞ design approach was partly stimulated by the need of the aerospace industry where robustness issues and safety aspects are of particular importance. The uncertainties in aircraft systems arise both from the variations in the system description at different operating conditions and the disturbances affecting the system.

Flight control system design for VSTOL aircraft is even more difficult than for usual applications. Control of the aircraft in manoeuvres from hover to wing borne flight pose particularly difficult multivariable highly interacting design problems. For an effective design the system must be properly analysed and appropriate models developed. The text introduces the basic components of the flight control and engine systems and describes the models for the different sub-systems.

An H_∞ loop shaping robust design procedure is described and the resulting system properties and results are analysed. Practical issues of controller implementation such as control switching and scheduling through different flight conditions are also considered. Problems of wind-up and the difficulties introduced by of the complexity of the solution are discussed. The resulting solution therefore satisfies both design requirements and can be implemented in practice.

In Part II the implementational problems are considered further, including the discretization process, handling limitations on actuators and the nonlinearities in the system. This text then goes further than most in being able to provide both piloted simulation and flight testing results.

The Monograph is of interest not only to flight control engineers but to designers in general industries which require a complete overview of procedures to be followed when applying one of the latest multivariable design techniques.

M.J. Grimble and M.A. Johnson
Industrial Control Centre
Glasgow, Scotland, U.K.

PREFACE

Recent developments in $\mathcal{H}_\infty$ theory have produced a highly attractive design approach. However, a large gap between theory and practice has emerged, there being as yet very few design examples applied to real industrial control problems. The work described in this monograph aims to narrow this gap, and to address implementation issues associated with multivariable $\mathcal{H}_\infty$ controllers. An $\mathcal{H}_\infty$ control law has been developed for the DRA Bedford research Harrier and is presented. This control law has recently undergone preliminary flight testing.

Future aircraft are likely to have many more control surfaces than current aircraft. To fully utilise these surfaces the pilot will have to rely increasingly on automatic control. Conventional design techniques do not provide a systematic way of designing for systems with multiple inputs and outputs. The paradigm of $\mathcal{H}_\infty$ is one way of directly designing for this type of system without the need for successive loop closing. The Harrier provides a particularly good design example in that it has both aerodynamic and vectored thrust controls which result in a 3-input system. Each of these three inputs has significant coupling to each of the controlled outputs. Thus demonstrating the $\mathcal{H}_\infty$ technology on the Harrier gives confidence that it could be applied to future aircraft for which a multivariable design approach may be necessary to extract maximum performance.

Early $\mathcal{H}_\infty$ design examples were based on optimisation using weighted closed-loop transfer functions. The control law presented here uses weighting of the open-loop frequency response to specify the desired performance, followed by an $\mathcal{H}_\infty$ robustness optimisation. The optimisation uses the normalised coprime factor uncertainty description. A weighting selection procedure is developed for this design approach. Then, given a set of linear designs which cover the flight envelope, the problem of moving from one design to the next as the flight envelope is traversed is addressed. Conventional controllers typically gain schedule proportional and integral gains. Two methods are considered for multivariable $\mathcal{H}_\infty$ controllers, one based on switching and one on gain interpolation of an observer implementation of the control law. Anti-windup schemes are also investigated, and a novel combination of using observers and the Hanus desaturation scheme is proposed.

A common criticism of $\mathcal{H}_\infty$ is that it produces high order controllers which are complex to implement. A design study has been carried out using parametric optimisation of a fixed structure controller in order to investigate the performance/complexity trade-off. The control law which was flight tested was model reduced using balanced truncation methods.

The control law developed for the DRA research Harrier XW175 is one of several control laws being evaluated by DRA Bedford under the Vectored thrust Aircraft Advanced flight Control (VAAC) programme. The VAAC Harrier and the DRA Large Motion Simulator (LMS) enable fast testing of new control law technologies. In the simulator the same non-linear model as used for design is flown with the control law by pilots. Once the control law is sufficiently mature, it can then be implemented on the VAAC Harrier.

The $\mathcal{H}_\infty$ control law gives full authority longitudinal control for airspeeds between the hover and 300 knots. Non-linear simulation, piloted simulation and flight test results are presented. The development of the control law is described in some detail, and it is intended that the text should have some tutorial value. It is hoped that the exposition of the control law in this monograph will help the reader apply $\mathcal{H}_\infty$ control to other real engineering applications.

Richard Alden Hyde
Girton College
Cambridge
28th December 1993

Acknowledgements

The material presented in this monograph is the result of five years research carried out at Cambridge University Engineering Department. For the first three years I was studying for my doctorate, and much of Part I originates from my thesis. For the past two years I have been a Research Associate, and all of Part II dates from this period. Funding was provided by the Science and Engineering Research Council, and the Defence Research Agency provided both simulation facilities and time on their XW175 research Harrier.

The direction of my work was greatly influenced by my PhD supervisor, Keith Glover. Keith's encouragement and advice have been invaluable, and the success of the project owes much to his support. I would also like to mention Steve Williams who originally motivated my interest in control when he supervised me as an undergraduate. It was Steve who set up the research programme between the DRA and Cambridge University. Glenn Vinnicombe, with whom I shared an office, imparted much helpful advice when I was designing for flight testing. All of my colleagues in the control group at Cambridge also deserve a special mention.

I would like to thank DRA Bedford for their involvement in the project, and providing us with access to their simulation facilities. They also provided us with a very rare opportunity, that of actually flight testing our control law design. The Deutsche Institut für Luft und Raumfahrt also very kindly gave me the opportunity to visit them for a few months and to try out their software on the VSTOL design.

CONTENTS

PART I

DESIGN AND IMPLEMENTATION OF H_∞ CONTROLLERS

CHAPTER 1
INTRODUCTION

Any model of a physical system will never be a true representation of that system. Often this is primarily because there are characteristics which cannot easily be modelled, and some approximation must be used. In any case, however detailed the model, there will always be modelling inaccuracies. In practice there is also a trade-off to be made between model complexity and model accuracy in that a simple model is easier to simulate and use for design. Robustness to modelling inaccuracy is therefore a vital characteristic required of a controller.

For single-input single-output controller designs, robustness is achieved by ensuring good gain and phase margins. For multivariable systems, gain and phase margin are no longer adequate, and measures of robustness using the $\mathcal{H}_\infty$-norm have been introduced [1]. Design using the "$\mathcal{H}_\infty$ approach" has recently become very attractive with the state-space closed form solution of [2]. This only requires the solution of two Riccati equations, and results in a controller of state dimension no higher than that of the plant. Earlier frequency domain methods resulted in much higher order controllers (see for example [3]). Because a multivariable $\mathcal{H}_\infty$ design approach addresses a priori the fact that a model of a plant will always be uncertain, it looks highly attractive for design.

Now that the optimisation machinery for the general $\mathcal{H}_\infty$ problem has been solved, there is a need for some realistic and comprehensive design studies using the approach to assess its applicability to real design problems. Part I of this monograph applies $\mathcal{H}_\infty$ techniques to a Generic VSTOL Aircraft Model (GVAM) supplied by the Defence Research Agency (DRA). The GVAM is a realistic model representative of current VSTOL aircraft, and as such presents a representative design problem on which to evaluate design techniques. The GVAM and controller can be piloted on the DRA simulator at Bedford for piloted evaluation. Most of what appears in Part I was developed in [4]. The discussion of flight control modes is new, and so is the anti-windup scheme which integrates the observer and Hanus desaturation schemes.

Part II uses the techniques developed in Part I to develop a flight control law suitable for flight testing on the DRA XW175 research Harrier. This control law is one of several control laws under evaluation at DRA Bedford, and has been designated "Control Law 005". VSTOL aircraft provide an excellent design example in that the basic aircraft is very highly coupled. The main coupling results from pitching moments generated by changes in thrust and nozzle settings. The vectored thrust also adds an extra input and an extra output, and thus makes a more challenging design problem as compared to a conventional aircraft. The control problems encountered when using limited authority vectored thrust to control high angles of attack could almost certainly be tackled using the same techniques as those developed here. The results of the design process and the final flight testing of control law 005 could therefore be highly relevant to future design of Integrated Flight and Propulsion Control Systems (IFPCS).

There are 16 chapters, the contents of which are now outlined.

PART I

Chapter 2 : The Motivation for Robust Control

This chapter briefly reviews the classical approach to robust control, and shows by example that gain and phase margins are not good indicators of robustness for multivariable systems. The paradigm of $\mathcal{H}_\infty$ is then motivated as a better framework for the description of uncertainty. Finally the structured singular value is briefly introduced.

Chapter 3 : A Multivariable Design Case Study

The GVAM is introduced, and operational requirements of VSTOL aircraft discussed. Previous studies on the GVAM are reviewed, and the context of the design study set. Handling quality requirements are reviewed, and interpreted in the context of multivariable control. Then the main nonlinearities of the GVAM are discussed, and ways of accounting for this when designing linear controllers are discussed. In particular, the need for some form of gain scheduling is motivated. Uncertainty modelling is then discussed, and finally a performance specification for initial linear feedback designs is given.

Chapter 4 : Robust Design Using Loop-Shaping

A procedure for selecting loop-shaping weights for the loop-shaping plus normalised coprime factor robust stabilisation design approach is developed. Specific attention is paid to ill-conditioned plants, and the importance of careful scaling when evaluating robustness measures is demonstrated. Some justification of the weights selection procedure is given, and it is shown how the procedure is easily configured to meet the aircraft handling quality requirements given in chapter 3. The procedure is also discussed in the light of work by Freudenberg [5] on loop-shaping. Finally some design examples are presented, including a hover design for the GVAM.

Chapter 5 : Controller Switching

In this chapter the idea of switching between linear controller designs as a form of scheduling is motivated. The approach of Hanus [6] is used to effect bumpless transfer between controllers on the GVAM. A conservative stability test is presented which indicates potential switching instabilities.

Chapter 6 : Controller Scheduling

This chapter shows how $\mathcal{H}_\infty$ controllers from the loop-shaping procedure may be smoothly gain scheduled with flight condition. The approach relies on the exact plant observer structure unique to the normalised coprime factor robust stabilisation approach. Two design examples on the GVAM are presented, and some analysis of the scheduled system carried out using the structured singular value.

Chapter 7 : Multivariable Anti-windup

The extra problems created by actuator windup for multivariable systems is discussed. Various anti-windup approaches are presented, and their relative merits discussed.

Chapter 8 : Controller Complexity

The complexity of the $\mathcal{H}_\infty$ controllers and associated switching/scheduling is much higher than that of conventional single loop controllers. The trade-off between complexity and performance for multivariable controllers is investigated using the Multi-Objective Programming System of the Deutsche Institut für Luft und Raumfahrt (DLR) which allows the parameters of a fixed structure controller to be optimised.

PART II

Chapters 9–16 : Development of a flight control law for the DRA research Harrier

A flight control law for testing the the DRA research Harrier XW175 is developed. The design makes use of many of the techniques developed in Part I. The linear design described in chapter 10 is carried out in more detail than the illustrative designs in Part I, in particular allowing for computer transport delays and sensor dynamics. Choice of feedback variables is also changed to suit the particular sensors available on the aircraft and sensor noise characteristics. The scheduling approach as opposed to switching approach is selected to give full flight envelope operation. Discretisation of the control law is covered in chapter 11, and direct discrete time design is used. Chapter 12 describes the non-linear command precompensation used external to the closed-loop to attain the desired flight modes and closed-loop responses. Chapter 13 covers other non-linear aspects including flight envelope limiting.

Chapters 14 and 15 give unpiloted and piloted simulation results, and finally conclusions are drawn in chapter 16.

Appendices

Appendix A: List of variable names.

Appendix B: Block diagram of control law 005.

CHAPTER 2
THE MOTIVATION FOR ROBUST CONTROL

2.1 Introduction

A model of a plant can never be perfect, and as such will always be an approximation to the true plant. Often certain characteristics of a plant will not be modelled at all. This is either because their contribution to the overall behaviour is small, or because they are not easily modelled. Furthermore, the plant's dynamics may change during long-term operation. To address the difference between modelled and true plants, various measures of robustness are used. A controller is said to exhibit good robust stability if it remains stable for all variations in plant behaviour which are reasonably expected to occur. Similarly a controller is said to exhibit good robust performance if it carries on performing satisfactorily for all encountered plant variations.

In classical single-input single-output (SISO) control, gain and phase margins are used as measures of robustness [7]. Loop shaping can be done in a very systematic way to attain good gain and phase margins, and to specify the desired closed-loop performance. If the plant has more than one output to control, then typically loops are closed sequentially i.e. a SISO controller is designed for one of the outputs using an appropriate actuator, that loop is then closed, and then another SISO controller is designed for the next output and so on. For example, for an aircraft control system the pitch attitude controller might be designed first using the tailplane, and then an airspeed controller designed using engine thrust for actuation.

It is well known that this sequential loop closing approach limits what performance and robustness levels can be achieved for many plants. In the next section this will be illustrated with a simple example, and the potential advantages of multivariable control motivated. This same example will then be used to illustrate the also well known fact that the quantities gain margin and phase margin are no longer reliable measures for robustness. Section 2.3 then discusses robustness in the $\mathcal{H}_\infty$-norm setting as replacement for gain and phase margin for multivariable plants, and the current $\mathcal{H}_\infty$ design synthesis approaches are reviewed. The conservativeness of the $\mathcal{H}_\infty$ approach is then discussed, and the structured singular value introduced in §2.4 which addresses this. Finally §2.5 summarises the main points of the chapter.

2.2 Multivariable Control

2.2.1 Motivation

The practice of closing successive loops with SISO controllers can unnecessarily limit the achievable behaviour of the closed-loop system. In particular coupling between controlled outputs may be unnecessarily large. To illustrate this consider the plant

$$G = \begin{bmatrix} \frac{1}{s+1} & \frac{1/2}{s+1} \\ 0 & \frac{1}{s+2} \end{bmatrix}$$

which has a high degree of coupling from the second actuator to the first output. In this simple design example the two SISO controllers are restricted to be proportional gains. The overall controller will be of the form

$$K = \begin{bmatrix} k_1 & 0 \\ 0 & k_2 \end{bmatrix}$$

if two successive SISO designs are performed. If the closed-loop is implemented as in Figure 2.1, then the transfer function from references to outputs will be

$$(I + GK)^{-1}GK = \begin{bmatrix} \frac{k_1}{s+1+k_1} & \frac{(s+2)k_2/2}{(s+1+k_1)(s+2+k_2)} \\ 0 & \frac{k_2}{s+2+k_2} \end{bmatrix}$$

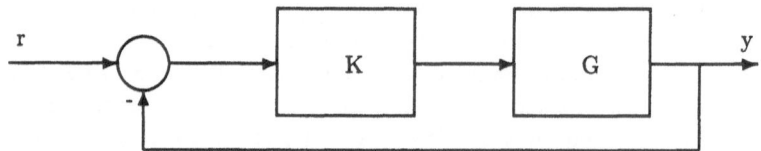

Fig. 2.1. Controller in the forward loop

Looking at the elements of $(I + GK)^{-1}GK$ it can be seen that there is nothing which can be done to reduce the coupling from demands on the second output to the first output if the system is to have a fast response; as the bandwidth of the second loop is pushed up by making k_2 large, the relative size of the coupling becomes larger. On the other hand, if a multivariable controller of the form

$$K = \begin{bmatrix} k_1 & k_3 \\ 0 & k_2 \end{bmatrix}$$

is used, then the closed-loop becomes

$$(I + GK)^{-1}GK = \begin{bmatrix} \frac{k_1}{s+1+k_1} & \frac{(s+2)(k_3+k_2/2)}{(s+1+k_1)(s+2+k_2)} \\ 0 & \frac{k_2}{s+2+k_2} \end{bmatrix}$$

Putting $k_3 = -k_2/2$ completely removes the coupling for the nominal system. Note that this does not affect the nominal pole locations of the closed-loop which are only affected by the choice of k_1 and k_2.

This clearly illustrates the potential benefit of better decoupling achievable with a multivariable design. It is also possible that better robustness levels may be achievable by allowing the controller to be multivariable. However, caution must be used when evaluating robustness of multivariable systems; it will be shown in the next section that classical gain and phase margins can be poor indicators of robustness.

2.2.2 Limitations of gain and phase margins

To extend the idea of gain and phase margins to multivariable systems the obvious thing to do is to break the individual loops one by one and evaluate the gain and phase margins for each of them. Breaking the loops individually for the simple example of the previous section at the plant output gives the transfer functions $\frac{k_1}{s+1}$ and $\frac{k_2}{s+2}$ for the individual loop transfer functions. Clearly these indicate infinite gain margin and 90^0 phase margin for all positive values of k_1 and k_2 which might suggest that the controller gives a very robust closed-loop. However, this is not necessarily the case. Consider a small modelling error $\frac{\epsilon}{s+2}$ in the (2,1)-element of the plant: the perturbed plant is then

$$G_\Delta = \begin{bmatrix} \frac{1}{s+1} & \frac{1/2}{s+1} \\ \frac{\epsilon}{s+2} & \frac{1}{s+2} \end{bmatrix}$$

It is easy to show that the closed-loop characteristic equation becomes

$$s^2 + (3 + k_1 + k_2 + \epsilon k_3)s + 2 + k_2 + \epsilon k_3 + 2k_1 + k_1 k_2 - k_1 k_2 \epsilon/2 = 0$$

If $k_3 >> k_1, k_2$, then a small negative ϵ can send the closed-loop unstable as $(3 + k_1 + k_2 + \epsilon k_3)$ can become negative. Hence the system may be very unrobust to practical uncertainty. Note that this was not indicated by the individual gain and phase margins of the two loops.

2.2.3 Extension of classical design to multivariable plants

The Nyquist criterion can be extended to the Generalised Nyquist criterion which was primarily developed by MacFarlane [8]. The criterion states that ([9][Theorem 2.9]) *if G(s) has p unstable poles, then the closed-loop system with return ratio −kG(s) is stable if and only if the characteristic loci of kG(s), taken together, encircle the -1 point p times anti-clockwise, assuming that there are no hidden unstable modes.* The characteristic loci are the loci of the eigenvalues of $G(j\omega)$ for $-\infty \le \omega < \infty$. Two design methodologies based on shaping the characteristic loci so as to satisfy the criterion were developed, and are called the Characteristic Locus Method, and the Direct Nyquist Array Method. A detailed account of these methodologies can be found in Maciejowski [9]. They have been applied by Williams [10] to a VS-TOL design example. Their main drawback is that they rely on the gain and phase margins of the characteristic loci, and as has already been discussed these can be poor measures of robust stability. Both approaches rely on the skill of the designer to shape the characteristic loci in order to satisfy the Nyquist criterion, and often the process is not straightforward.

2.3 Design in $\mathcal{H}_\infty$

2.3.1 Using singular values as a robustness measure

Extension of gain margin and phase margin to multivariable systems is a poor indicator of robustness primarily because it does not allow for coupling between loops. An alternative uncertainty description which does allow for coupling makes use of the maximum singular value of various closed-loop transfer functions to quantify robustness levels. The maximum singular value of some matrix M is denoted as $\bar{\sigma}(M)$, and is defined as

$$\bar{\sigma} := \max_{u, \|u\| < 1} \|M.u\|$$

where $\|x\|$ denotes the Euclidean vector norm defined as $\sqrt{(x^H x)}$, and u is an input vector of appropriate dimension. In effect the maximum singular value is the maximum vector gain over all input directions.

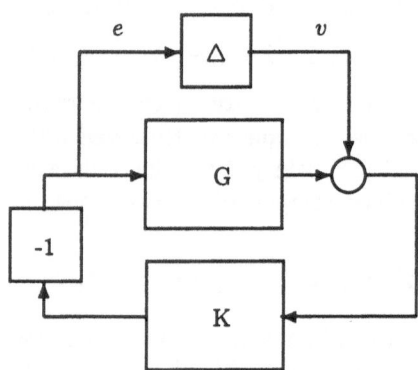

Fig. 2.2. Unstructured additive uncertainty.

Consider Figure 2.2. The Δ-block represents additive uncertainty associated with the nominal plant G, and as such will be a function of frequency. Assume that the closed-loop formed by G and K is stable for $\Delta = 0$. Now consider a non-zero Δ. If it is sufficiently large such that the gain measured by breaking the loop at v is greater than one, then instability may result (this can be formally prooved by application of the Small Gain Theorem [11],[12] which states that a necessary condition for instability is that the loop gain is greater than one).

Hence given an upper bound on the maximum singular value of $\Delta(j\omega)$

$$\overline{\sigma}[\Delta(j\omega)] < \epsilon$$

the closed loop is guaranteed to remain stable if the transfer function from v to e is less than $\gamma := 1/\epsilon$ i.e.

$$\overline{\sigma}[K(I + GK)^{-1}] < \gamma \quad \text{for all} \quad \omega$$

Other uncertainty descriptions are possible. Figure 2.3 illustrates multiplicative output uncertainty. In this case, if the nominal closed-loop is stable and

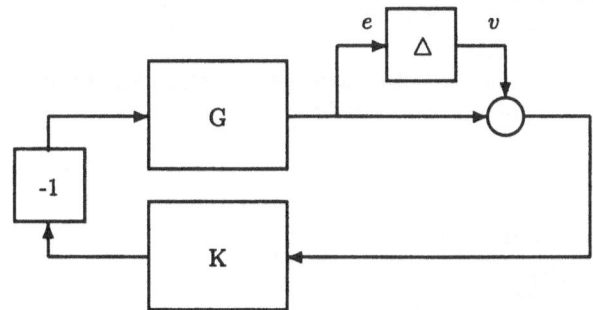

Fig. 2.3. Unstructured multiplicative output uncertainty.

$\overline{\sigma}[\Delta] < \epsilon$, then by the small gain theorem the perturbed system is guaranteed stable if

$$\overline{\sigma}[GK(I + GK)^{-1}] < \gamma \quad \text{for all } \omega$$

where $\gamma := 1/\epsilon$. A design objective might be to find a K which maximises the size of ϵ for which the closed-loop remains stable for a particular uncertainty description. To achieve this goal the H_∞ norm defined as

$$\|G\|_\infty := \sup_\omega \overline{\sigma}[G(j\omega)]$$

is used. Hence the H_∞ norm is the maximum singular value over all frequencies. In H_∞ control law synthesis, the K which minimises the H_∞ norm of a collection of transfer functions is found. The chosen set of transfer functions to optimise is put in to the so-called "standard form" which is defined in the next section. Once in this form, the optimal controller is found using standard software e.g. [13]. Two examples of $\mathcal{H}_\infty$ design objectives are introduced in §2.3.3 and §2.3.4. For the reader unfamiliar with $\mathcal{H}_\infty$ further introductory material can be found in [9] and [14].

2.3.2 The $\mathcal{H}_\infty$ standard plant

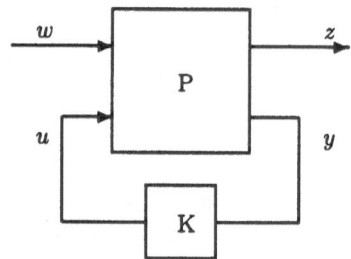

Fig. 2.4. The $\mathcal{H}_\infty$ standard plant.

All combinations of transfer function objectives can be formulated into the standard form of figure 2.4 where P is partitioned as

$$P = \begin{bmatrix} P_{11} & P_{12} \\ P_{21} & P_{22} \end{bmatrix}$$

The objective is to find K which achieves

$$\inf_{\text{stabilising } K} \|\mathcal{F}_L(P, K)\|_\infty$$

where $\mathcal{F}_L(P, K)$ denotes the lower Linear Fractional Transformation (LFT) defined by

$$P_{11} + P_{12}K(I - P_{22}K)^{-1}P_{21}$$

i.e. the transfer function from w to z. Figure 2.5 shows the additive uncertainty description of Figure 2.2 topologically transformed into the standard form. Once in this form, standard software tools (e.g. [13]) can be used to find the optimal controller K.

The theory of $\mathcal{H}_\infty$ optimal control originated with Zames in 1981 [1]. The solution techniques at the time only effectively addressed the SISO case as the frequency domain solution methods ran into technical problems for the multivariable case. The later state-space solution (see for example Francis [3]) were able to address the multivariable case, but resulted in controllers of high state dimension. A recent solution to the standard problem is that of [2], [15], and is referred to as the DGKF solution. This approach involves solving only two Riccati equations, and results in a controller of state dimension no

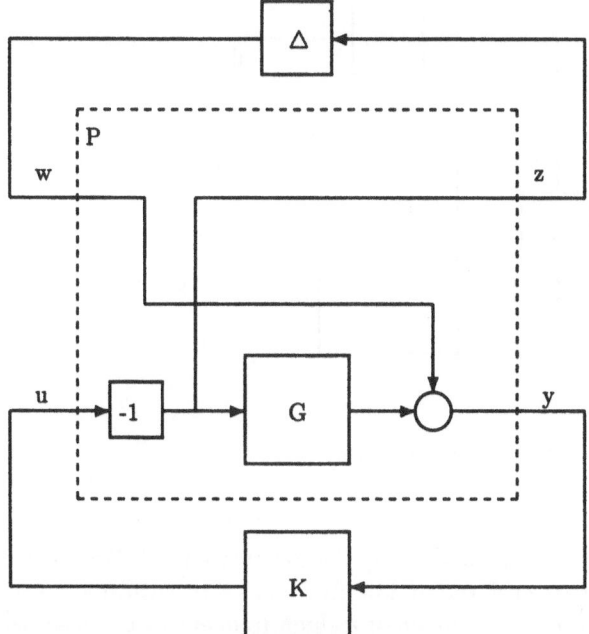

Fig. 2.5. The $\mathcal{H}_\infty$ standard plant for additive uncertainty.

higher than that of the weighted plant. The advent of this reliable solution technique has made the $\mathcal{H}_\infty$ problem formulation a very attractive design approach. As the emphasis of this monograph is that of the practicalities of applying $\mathcal{H}_\infty$ control to a realistic design example, the controller equations are not presented here. For these the reader is referred to [15].

2.3.3 The S & KS design procedure

The $\mathcal{H}_\infty$ minimisation problem illustrated in Figure 2.5 does not represent a sensible practical design problem. Consider, for example, the case where G is stable. In which case making $K = 0$ gives closed loop stability for any size of Δ. However, making $K = 0$ gives zero feedback and hence no disturbance rejection. To make a sensible problem, a performance objective must also be included within the standard plant P.

Consider Figure 2.6. The additive uncertainty path corresponds to the transfer function from d to e_2. The additional path from d to e_1 corresponds to the sensitivity i.e. the output disturbance rejection. Hence minimising the $\mathcal{H}_\infty$ norm of the transfer function from d to $\begin{bmatrix} e_1 \\ e_2 \end{bmatrix}$ attempts to maximise robustness to additive uncertainty and to maximise the rejection of output disturbances. There is of course a conflict here in that maximising disturbance rejection implies high gains which is at odds with good robustness. Hence

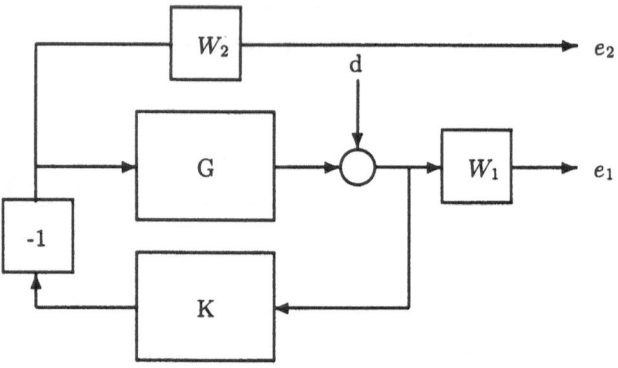

Fig. 2.6. S & KS design formulation

the weighting functions W_1 and W_2 are introduced. The sensitivity weight W_1 is chosen to be large at low frequencies to ensure good low frequency disturbance rejection, and small at high frequencies to reflect that the plant is not physically capable of rejecting high frequency disturbances. Conversely the additive uncertainty weight is made small at low frequencies, and large at high frequencies where unmodelled dynamics make the plant model more uncertain. The designer is therefore making a robustness/performance trade-off when selecting the weights.

Stated more mathematically, the design goal is to find K which achieves the infimum of

$$\inf_{\text{stabilising } K} \left\| \begin{bmatrix} W_1 \\ W_2 K \end{bmatrix} (I + GK)^{-1} \right\|_\infty$$

The formulation is referred to as the "S and KS" approach where $S = (I + GK)^{-1}$ is the sensitivity. The transfer function from d to e_2 can also be interpreted as putting a penalty on actuator gain. As for the additive uncertainty interpretation, W_2 needs to be made large at high frequency so as to prevent high frequency actuator use.

It has been shown [16],[17] that the controller from the $S\&KS$ formulation exactly cancels plant poles. This provides a particular problem if the plant has lightly damped poles in that cancelling them will result in a very unrobust system. This was found to be the case for the VSTOL designs in [18].

A physical interpretation of the pole cancellation is that the disturbance only acts at one point in the loop, namely at the plant output. The optimisation expoits the fact that there is no disturbance on the plant input which can excite plant poles, and cancels the signal path from the output disturbance to the plant input with zeros.

Adding an extra disturbance at the plant input, and keeping the same two errors e_1 and e_2 results in the following optimisation:

$$\inf_{\text{stabilising } K} \left\| \begin{bmatrix} W_1 \\ W_2 K \end{bmatrix} S[W_3 \ GW_4] \right\|_\infty$$

This optimisation problem does not have the pole-zero cancellation problem. However, the selection of the weights is much harder in that four weights must now be selected.

2.3.4 The loop-shaping/coprime factor uncertainty approach

This approach was developed by Glover and McFarlane [19],[20] and makes use of an uncertainty description based on additive perturbations to a normalised coprime factorisation of the plant. A normalised left coprime factorisation (LCF) of a plant G is given by $G = \tilde{M}^{-1}\tilde{N}$ where $\tilde{N}$ and $\tilde{M}$ are coprime matrices in $R\mathcal{H}_\infty$ and

$$\tilde{N}\tilde{N}^* + \tilde{M}\tilde{M}^* = I$$

Robustness with respect to additive perturbations to the normalised coprime factors is considered. The class of perturbed system models is given by the family

$$\mathcal{G}_\epsilon = \left\{ (\tilde{M} + \Delta_M)^{-1}(\tilde{N} + \Delta_N) : [\Delta_M, \Delta_N] \in R\mathcal{H}_\infty^{p\times(p+m)}, \|\Delta_M, \Delta_N\|_\infty < \epsilon \right\}$$

The problem formulated and solved in [19] is to find the largest class of such systems, i.e. the maximum of ϵ, $\epsilon_{\max}$, such that $\mathcal{G}_{\epsilon_{\max}}$ can be stabilised by a single fixed controller, K. Notice that this uncertainty framework allows the number of right half plane poles of the perturbed plant to be different to that for the nominal plant. If the plant has poles near the imaginary axis, this uncertainty description seems more appropriate than those used in the standard closed-loop $\mathcal{H}_\infty$ objectives which do not allow poles to cross the axis. The solution is given by minimising

$$\epsilon_{\max}^{-1} = \gamma_{\min} = \inf_K \left\| \begin{bmatrix} K \\ I \end{bmatrix} (I - GK)^{-1} \tilde{M}^{-1} \right\|_\infty$$

The solution to this problem is particularly attractive in that the optimal γ can be found without recourse to the γ-iteration which is normally required to solve $\mathcal{H}_\infty$ problems. Given a minimal realisation $[A, B, C, 0]$ of a controllable and observable plant, the Control Algebraic Riccati Equation (CARE) and Filtering Algebraic Riccati Equation (FARE) are given by

$$A^*X + XA - XBB^*X + C^*C = 0$$

$$AZ + ZA^* - ZC^*CZ + BB^* = 0$$

The optimal γ is simply given by

$$\gamma_{\min} = (1 + \lambda_{\max}(XZ))^{1/2}$$

and the central controller (see [19],[21]) is given by

$$K = \left[\begin{array}{c|c} A + HC + \gamma^2 BB^* XW_1^{*-1} & -H \\ \hline \gamma^2 B^* XW_1^{*-1} & 0 \end{array} \right]$$

where

$$W_1 := I + XZ - \gamma^2 I$$

and

$$H = -ZC^*$$

As given so far the robust coprime factor stabilisation procedure addresses robustness, but does not give the designer a way of specifying performance. To specify performance McFarlane and Glover propose pre- and post-compensating the plant to give the desired open-loop singular values. The shaped plant is thus

$$G_s = W_2 G W_1$$

where W_1 and W_2 are pre- and post-compensators respectively. This loop-shaping procedure is a very natural extension to classical loop shaping for SISO plants. The whole of chapter 4 is devoted to the selection of weighting functions for multivariable plants, and this loop-shaping approach is used for the designs in Part II.

2.4 The structured singular value

The use of the maximum singular value as a norm on unstructured uncertainty overcomes the limitations of classical gain and phase margins by allowing for coupling between loops. However, often in reality the uncertainty may be highly structured, and ignoring this can lead to very conservative results. For example, input multiplicative robustness might be used for modelling actuator uncertainty, and in which case the uncertainty block will have the structure

$$\Delta = \left[\begin{array}{cccc} \delta_1 & & & \\ & \ddots & & \\ & & \ddots & \\ & & & \delta_u \end{array} \right]$$

for u actuators.

To build such knowledge about the uncertainty structure into analysis and design, the structured singular value (sometimes referred to as μ) [22],[23] can be used. Figure 2.7 illustrates the general set-up. The Δ can be used to

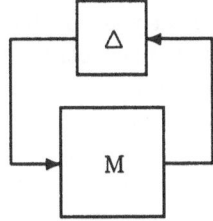

Fig. 2.7. Set-up for the structured singular value test.

represent both uncertainty and performance, and has the general structure

$$\Delta = \{\text{diag}[\delta_1 I_{r_1}, ..., \delta_s I_{r_s}, \Delta_1, ..., \Delta_F] : \delta_i \in C, \Delta_j \in C^{m_j \times m_j}\} \subset C^{n \times n}$$

M contains the interconnection of the controller and plant, and is assumed nominally stable. The structured singular value of M is defined as

$$\mu_\Delta(M) := \frac{1}{\min\{\overline{\sigma}(\Delta) : \Delta \in \Delta, \det(I + M\Delta) = 0\}}$$

unless no $\Delta \in \Delta$ makes $I + M\Delta$ singular, in which case $\mu_\Delta(M) = 0$. Hence $\mu_\Delta(M)$ is a measure of the smallest structured Δ that causes instability. Notice that the maximum singular value of M is an overbound in $\mu(M)$, and as such is conservative. Comprehensive software is now available which solves μ for general complex or mixed complex and non-repeated real blocks [24],[13]. Structured singular value analysis is made use of in chapter 6 for analysis of scheduled systems.

Approximate synthesis is also possible with the structured singular value using the so-called D-K iteration. It makes use of the following upper bound on $\mu(M)$:

$$\mu_\Delta(M) \leq \inf_{D \in \mathbf{D}} \overline{\sigma}(DMD^{-1})$$

where

$$\mathbf{D} = \{\text{diag}[D_1, ..., D_s, d_1 I_{m_1}, ..., d_{F-1} I_{m_{F-1}}, I_{m_F}] :$$
$$D_i \in C^{r_i \times r_i}, D_i = D_i^* > 0, d_j \in R, d_j > 0\}$$

Essentially the procedure comprises of carrying out an initial $\mathcal{H}_\infty$ design, carrying out a μ-analysis, fitting optimal D's with rational stable and invertible functions which are then used as weighting functions for the next $\mathcal{H}_\infty$ design. This procedure is repeated until μ reaches a minimum. Further details can be found in [23], and design examples in [25],[26], and [18].

2.5 Summary

In this chapter the potential advantages of multivariable controllers over classical SISO controllers have been motivated, in particular the potential for better decoupling has been emphasised. It has also been shown that the classical measures of robustness, namely gain and phase margins, are no longer reliable indicators for multivariable systems. It has been argued that the design methods which are natural developments of classical SISO control, and which rely on shaping the characteristic loci to satisfy the generalised Nyquist criterion may lead to poor robustness. The idea of robustness to unstructured uncertainty in the $\mathcal{H}_\infty$ norm was introduced, and it was argued that this is more appropriate as it allows for coupling between loops. The combination of multivariable and $\mathcal{H}_\infty$ techniques give the potential for achieving good performance along with good robustness properties with a straightforward design procedure.

The main emphasis of this monograph is to apply robust multivariable control to a realistic design example, and to evaluate the potential benefits and the practicalities of the approach. Specifically the loop-shaping/coprime factor robust stabilisation approach of McFarlane and Glover is concentrated on. It it considered that this approach has a more intuitive weights selection than that of the S & KS approach, and initial design studies showed it more suited to the design study presented here, particularly for its observed robust performance properties.

CHAPTER 3
A MULTIVARIABLE DESIGN CASE STUDY

3.1 An introduction to VSTOL aircraft

Undoubtedly the best known fixed wing Vertical and/or Short Take-Off and Landing (VSTOL) aircraft is the Harrier jump-jet. Work on this type of aircraft began in the fifties at Bedford with the "flying bedstead" experiments. The present-day GR7, FRS.2 and AV-8B Harrier are essentially upgraded versions of the original P1127 Harrier which first flew in 1960, the main differences being increased operating range, much advanced avionics and improved handling qualities. The vectored thrust capability has proved itself to have unique strategic advantages not only through short take-off and landing capability, but also for high manoeuvrability in the air. In addition to the obvious advantages of VSTOL operations from small destroyers at sea [27], they have also been used to great effect by the U.S. Marine Corps for their ability to support and supply the up-front ground troops [28].

The handling qualities of even the modern-day Harrier versions impose a high workload on the pilot, particularly during VSTOL operations. Control is particularly hard in the absence of outside motion cues, and in gusty weather conditions which can severely limit shipboard operations [27]. The high pilot workload in controlling the aircraft can be seen as the result of three main factors:

1. Poor inherent stability. The aircraft is longitudinally unstable at low speed, and so the pilot has to make continuous corrective actions to hold the attitude. It has a divergent response in roll which means the pilot must ensure that he doesn't put on too much bank angle. In yaw there is very low directional stability, and high yaw rates at low speeds are potentially lethal [29]

2. Inceptor configuration. The pilot has to operate both the nozzle lever and throttle setting with his left hand, leaving the right hand free to control the aircraft in pitch and roll using the stick. During a typical descent the pilot has to step the nozzle lever through 20, 40, 65 and finally 80 degrees with appropriate increases in the throttle setting [29].

3. Poor decoupling. The pilot's inceptors do not command directly what the pilot wishes to control; for example, a change in nozzle lever setting alters both vertical and horizontal accelerations.

Automatic control has great potential for addressing all three of these undesirable characteristics. The present-day Harriers have various control systems to improve handling qualities, but only really address the first of the three above causes of poor behaviour. There is potential to address all three causes with multivariable control as it enables decoupling between demands as was discussed in the previous chapter.

A Generic VSTOL Aircraft Model (GVAM) is used in Part I to investigate the applicability of multivariable robust control to this type of aircraft. An introduction to the GVAM used for the study is given in §3.2, and a review of present aircraft control systems and previous design studies on the GVAM are given in §3.3. The rest of the chapter is associated with arriving at the design specification in §3.7 for linear multivariable controller designs. To achieve this goal, §3.4 reviews handling qualities criteria, §3.5 the plant's key non-linearities, and §3.6 discusses modelling uncertainty. Issues concerned with specifications for command prefilters and non-linear behaviour associated with large demands are dealt with in Part II alongside a detailed control law design.

3.2 The Generic VSTOL Aircraft Model (GVAM)

The Generic VSTOL Aircraft Model (GVAM) was developed by the Royal Aerospace Establishment, RAE, (now the Defence Research Agency, DRA) to provide a vectored thrust aircraft model for use in Advanced Short Take-Off and Vertical Landing (ASTOVL) control law studies, and for real-time piloted simulation [30]. The model is highly comprehensive, taking account of such things as aerodynamic interference effects from the vectored thrust as well as modelling the aerodynamic coefficients and engine non-linear effects. As such it provides a realistic design problem, and captures most of the characteristics which would be encountered when designing for a real VSTOL aircraft. Some of the more important non-linearities which impact directly on the control law design are discussed in §3.5.

The GVAM is configured for use with the TSIM package for simulation and extraction of linearisations for design work. The TSIM interface makes easy access to variables in the model. Figure 3.1 illustrates the GVAM in transition from powered lift to wing-borne flight.

As with any conventional aircraft the longitudinal equations of motion of the airframe can be represented in a linearised form using four states, namely pitch attitude, pitch rate, and the two longitudinal velocities, {THETR,Q,UB,WB}. These four states give rise to the aircraft short period and phugoid modes; essentially it is the damping and frequency of the short period mode which dictates the aircraft's pitch response following a disturbance or tailplane demand. A derivation of the equations of motion for conventional wing-borne aircraft can be found in Babister [31] or McLean [32]. A complete linearisation for aircraft longitudinal input-output behaviour

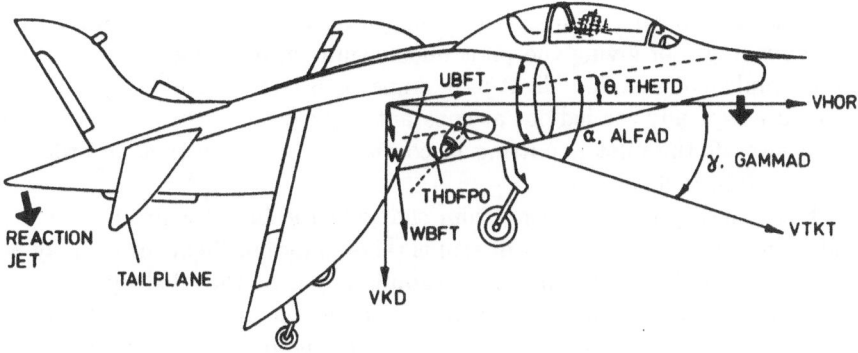

Fig. 3.1. GVAM in transition to wing-borne flight

will also have actuator, sensor and engine states in addition to the airframe states. Linearisations for the lateral motion can also be obtained from the GVAM, but in this design study only control of the longitudinal motion is considered.

Control laws can then be developed using linear design approaches, thus taking advantage of the highly developed linear optimal control theory. Linear controller designs can then be evaluated on the GVAM using the TSIM package. Following this it is possible to evaluate them in real-time using the DRA piloted simulation facilities at Bedford.

3.3 Flight control modes

In its simplest form, Fly-by-Wire (FBW) is just used to remove heavy mechanical and hydraulic control runs and thus reduce aircraft weight. However, the full advantages of FBW are only realised when the direct links from the conventional cockpit controls to the aerodynamic surfaces are broken and routed via a flight computer. With this system, the flight computer reads in the pilot's demands and then makes appropriate changes to the control surfaces. This gives rise to a whole host of possibilities, including the use of feedback to change the flying qualities of the aircraft. An example might be to change the short period and phugoid modes of the controlled aircraft so that they give better flying qualities or even emulate those of another aircraft. With such systems the pilot is still using the same cockpit controls which still have the same functionality as those of the uncontrolled aircraft. It is, however, possible to take it one step further, and to change the functionality of the cockpit levers or even give the pilot a completely new set of control levers. Such an example would be to give the pilot a flight path angle demand on a side stick instead of the conventional elevator demand from centre stick. The advantages of giving the pilot unconventional demands are numerous. For example, commanding flight path angle leads to easy landing approaches which are typically carried out at constant flight path angle; with a conventional aircraft the pilot is putting in continual stick corrections in order to hold the desired flight path.

A potential problem arising from changing the inceptor functionality is that a particular set of cockpit demands tailored for one flight regime may be totally inadequate for another. An example might be that flight path angle is appropriate for the landing approach, whereas commanding pitch rate is preferred during the landing flare as this gives the pilot precise control over the landing attitude. This necessitates different "control modes" across the flight envelope, and this has two important consequences. Firstly, the question of how and when to initiate modes must be tackled - and in particular how to indicate clearly to the pilot which modes he is in. Secondly, the demands for the mode which is about to come on line must be correctly initialised either directly by the pilot, or by the flight computer.

A potential problem with changed inceptor functionality is so-called "reversion to type", which essentially means that the pilot reverts to using the cockpit inceptors as if he were flying the conventionally controlled aircraft. This typically happens when there is a sudden increase in pilot workload, and can be potentially dangerous. An example on a Harrier would be giving the pilot a horizontal acceleration demand on the left-hand inceptor and vertical acceleration demand on the right-hand inceptor across the flight envelope; if when performing a vertical landing the pilot suddenly had to abort, he might be tempted to push fully forward on the left-hand inceptor corresponding to a thrust demand on a conventionally flown Harrier instead of pulling back on the right-hand inceptor.

The following broad philosophy was used for the development of flight control law 005. It is in no way claimed that this is a "best" philosophy, but it is used in an attempt to justify the choices made when specifying control modes.

- **P1** Keep the pilot in touch with the aircraft - don't remove him unnecessarily far from the physical system e.g. if the aircraft has a force motivator, give him an acceleration demand rather than a position or velocity demand. This way the aircraft "state" changes only when the pilot makes a change in the controls. An aircraft height demand system, for example, would not operate like this; the pilot would set a height, and then a while later when the height is achieved the pilot would perceive sudden changes in attitude and thrust.

- **P2** Avoid reducing the operating envelope of the aircraft; for example, if the pilot knows he can satisfactorily fly the aircraft manually up to 20 degrees angle of incidence at a given airspeed, he is not going to be impressed with a control system which only lets him fly up to 15 degrees incidence. If possible, extend the operating envelope of the aircraft if this gives rise to any mission benefits e.g. greater manoeuvrability.
- **P3** Don't restrict the pilot's authority to handle control redundancy, and ensure that it is handled in a natural way. For example, with a conventionally flown Harrier it is possible to achieve a given airspeed and descent rate with a variety of thrust, nozzle angle, tailplane and flap settings. The actual combination selected by the pilot might depend on factors such as how much surplus fuel he has - if it's low, then he'll aim to get as much aerodynamic lift as possible necessitating a high alpha. If fuel isn't a problem he may wish to have a low alpha giving him a lower pitch attitude and hence better view out of the cockpit. A control law which restricts this freedom is likely to attract criticism.
- **P4** In as far as is possible, attempt to get the directional functionality of the inceptors as close their functionality on a manually flown aircraft. This should minimise the risks if the pilot reverts to type demanding (see discussion above).
- **P5** Keep the number of command modes needed to cover the whole operating envelope to a minimum, and ensure that it is clear which mode the flight controller is in at any time. The higher the number of modes, the higher the chances become of incorrect mode selection and initialisation. A high number of modes also increases the overall complexity of the flight controller, and hence the amount of pilot training required is increased.
- **P6** Capitalise as much as possible on the benefits of multivariable feedback i.e. decoupled demands and disturbance rejection. Ensure that if the pilot lets go of all the controls that the aircraft is stable in the sense that airspeed, flight path and pitch attitude (if using vectored thrust) tend to constant mean values.

As will become clear in the following discussion of the inceptor functionality specification, the above desirable characteristics often have to be traded off against each other. This trade-off is particularly difficult for VSTOL aircraft since the vectored thrust gives rise to many more control possibilities than for conventional aircraft (see for example [29, Gainza]). By its nature the vectored thrust makes the manually flown aircraft very difficult to fly to its full potential, and hence the benefits of automatic control can be high.

3.4 Handling qualities

3.4.1 Introduction

Aircraft handling qualities can be defined as a measure of the ease with which a pilot can carry out specific tasks [32]. They depend not only on the frequency and damping of the aircraft's modes, but also on the control systems, inceptor configuration and aircraft instrumentation. There are a number of different criteria which have been developed to specify and evaluate handling qualities. Franklin [27] gives an excellent overview of VSTOL specific operational requirements, and discusses how these affect required handling qualities. The main handling qualities used for the design study here are U.S. MIL-Specification 83300 [33], U.S. MIL-Specification 8785C [34] and an Advisory Group for Aerospace Research and Development (AGARD) [35] report. MIL-83300 and the AGARD report give handling qualities specific to to VSTOL operations, whereas MIL-8785C is for fully wing-borne flight. These specifications are only an outline of required handling qualities, and are only for fully visual flight and calm air conditions. They are not specifically aimed at aircraft with automatic control systems, and in particular are not aimed at multivariable controllers.

More recently alternative handling quality metrics have been published [36],[37], and are referred to here as the Gibson Criteria. These criteria are much more suited to evaluate automatic control systems, and are partly based on time response characteristics. They can also predict potential Pilot Induced Oscillation (PIO) problems; the pilot in the loop can actually destabilise the aircraft if the control system has poor phase characteristics. The Gibson criteria are also used to evaluate handling qualities of the designs on the GVAM here.

In §3.4.2 the MIL-specifications and the AGARD requirements are summarised, and an attempt is made to extend them for use in a multivariable closed-loop control setting. Section 3.4.3 then outlines the Gibson criteria.

3.4.2 The AGARD and MIL–specifications

Tables 3.1.a and 3.1.b summarise the longitudinal handling quality requirements from the AGARD report and MIL-83300 document for Hover and Short Take-Off and Landing (STOL) operations. A discussion of requirements for fully wing-borne operation in MIL-8785C has been omitted as these result in similar guidelines for the closed-loop controller design, and also the Gibson criteria give a more up to date account of what is required at these speeds. The the numerical values in the table from MIL-83300 are for what is known as Level 1 handling qualities. Level 1 is defined as "Flying qualities adequate to accomplish the mission flight phase", and is more stringent than levels 2 and 3. Values given in the AGARD report are manoeuvre dependent, and the most demanding requirements have been summarised here.

Table 3.1.a – VSTOL pitch axis handling qualities				
Parameter	Demand Type	Minimum level		Source
		Hover	STOL	
1. Pitch angular acceleration (rad/sec)	Attitude	0.1–0.3	–	AGARD
	Rate	0.1–0.3	0.05–0.2	
	Acceleration	0.2–0.4	–	
2. Pitch angle after 1 sec (degrees)	Attitude	–	–	AGARD
	Rate	2–4	2–4	
	Acceleration	2–4	2–4	
3. Time to reach 63% peak angular acceleration, sec	Attitude	< 0.2	< 0.3	AGARD
	Rate	< 0.2	< 0.3	AGARD
4. Time to reach 90% de-manded atti-tude change, sec	Attitude	> 1 and < 2	—	AGARD
5. Time to reach max an-gular acceler-ation, sec	All	< 0.3		MIL-83300
6. Time for ac-celeration in commanded direction, sec	All	< 0.1		MIL-83300
7. Angular velocity damping, 1/sec	Attitude	-2.0	-1.0	AGARD
	Rate	-0.5 – -2.0	-1.0	
8. Roots of short period response $s^2 + 2\zeta\omega_n s + \omega_n^2 = 0$		$\omega_n > 0.35 rads^{-1}$ $2\zeta\omega_n > 1.0 s^{-1}$ $\zeta > 0.3$		MIL-83300
9. Normal ac-celeration due to sus-tained resid-ual oscillations		< 0.05g		MIL-83300

Table 3.1.b – VSTOL horizontal and vertical handling qualities		
Parameter	Minimum level	Source
10. Vertical acceleration command authority	+/- 0.1g	AGARD
11. Flight path/climb rate authority	$6°$ or 600feet/min	AGARD
12. Time to reach min vertical authority requirements	< 0.5 sec	AGARD
13. Time to reach 63% commanded thrust of at least 0.05W	< 0.3 sec	MIL-83300

Notice that the requirements for pitch response are much more comprehensive, and this reflects that this contributes most to pilot workload in the hover. The pitch requirements are the easiest to interpret for automatic control which is to be expected as pitch stability augmentation systems were already common practice when these handling qualities were drawn up. The horizontal and vertical requirements are heavily geared towards thrust response times and authority limits. However, deductions about horizontal and vertical motion demand responses can be made; for example, in the hover with the nozzles fully down, the requirement on thrust can be interpreted directly as a requirement on vertical acceleration demands.

Consideration of the data in the two tables leads to the following guidelines for designing the closed-loop controller:

1. If the aircraft actuators are utilised fully, then requirements 1,10 and 11 are likely to be satisfied if the basic aircraft already has sufficient control authority. This just requires that the pilot commands to the control system are suitably scaled, and as such does not directly affect the closed-loop controller design.
2. Provided the above is satisfied, and lags between demands and outputs are not too large, then 2,3,4,5,6,12 and 13 will be satisfied. In addition, 12 and 13 require that the lag in the forward loop between demands and actuators is small.
3. If the time responses to steps on pilot demands have little of no overshoot, no residual oscillations, and are sufficiently fast, then 7,8, and 9 can be satisfied.

Note that the fast response requirements do not necessarily imply that the closed-loop bandwidth must be high since a demand precompensator can be used to push up the effective bandwidth from demands to outputs. However, a closed-loop controller which meets handling qualities without complex precompensators has advantages in terms of easier implementation.

In chapter 4 it will be shown how the $\mathcal{H}_\infty$/loop-shaping design procedure is used to comply with these guidelines in a straightforward manner.

3.4.3 The Gibson criteria

An outline of the British Aerospace/Gibson criteria appeared in [36]. The criteria are primarily aimed at single loop control of pitch attitude with a pitch rate demand, and for conventional aircraft in fully wing-borne flight. However, the ideas can easily be extended to other flying tasks.

The Gibson criteria attach much more importance to the time-response rather than complex-plane pole locations. Figure 3.2 illustrates a typical response to a pilot pitch rate demand. Gibson states that the MIL-specifications

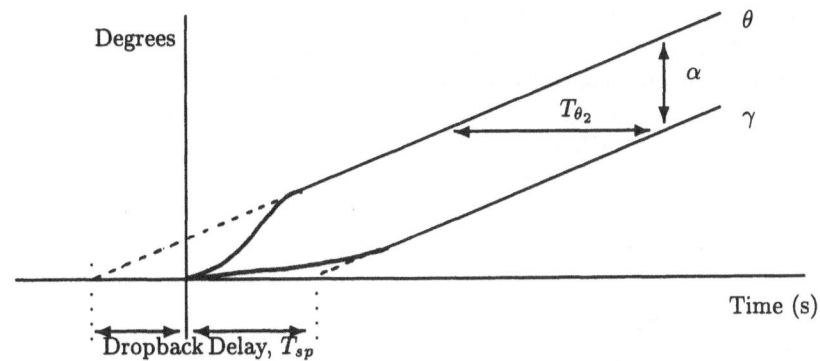

Fig. 3.2. Gibson time-response criteria.

are equivalent to bounds on the short period frequency, ω_{sp}, the short period damping, ζ_{sp}, and the flight path to attitude lag time constant, T_{θ_2}. This is with the exception of the MIL-requirements for (i) the time to reach peak acceleration, and (ii) the effective time-delay of pitch rate. Gibson replaces these with high frequency phase conditions.

For flight path tracking the quantities T_{θ_2} and the delay T_{sp} are most important. T_{sp} is dictated by the short period frequency, ω_{sp}. Gibson gives the quantity of about 1 second as an upper limit for satisfactory performance. For attitude control zero dropback is ideal.

Pilot Induced Oscillation (PIO) has been related to the rate of change of phase lag at the point where the phase between pilot demand and pitch output reaches $180°$; a too high phase rate is likely to give PIO problems.

There are also constraints on the attitude gain and the minimum frequency at the 180° phase point.

Gibson notes that the conventional aircraft has direct terms from the pilot controls to actuator demands, and that a control system which does not have direct terms in the forward path between references and actuators may well exhibit poor phase characteristics.

3.5 GVAM non-linearities

3.5.1 Introduction

VSTOL aircraft by their very nature are highly non-linear in operation. The non-linearity arises mainly from aerodynamic effects, the motivator mechanics and geometry, and the non-linear nature of the engine. However, with an understanding of the main non-linearities it is possible to successfully use linear design techniques in the development of a control law. A design approach which incorporates as much knowledge about the main non-linearities as possible into the control law, rather than leaving it for the robustness of the controller to cope with is clearly very desirable. In practice there is often a trade-off to be made here between how much knowledge is incorporated into the control law, and keeping controller complexity within reasonable limits. In the following sections, the main non-linear effects of the GVAM and the measures that were taken to account for them are given.

3.5.2 Key parameters

The term "key parameters" is introduced here to denote variables which capture non-linear effects. The aerodynamic coefficients which appear in the state-equations for the longitudinal motion of the aircraft depend on Mach number, Reynolds number and the angle of attack (see for example [31]). In the design study presented here, the key parameters used are airspeed (which is related to Mach number) and angle of attack. Linearisations for different discrete values of airspeed and incidence are made, and individual control designs carried out for each one. The linear controllers can then be switched or scheduled as a function of the key parameters using the techniques described in chapters 5 and 6. In this way the non-linear nature captured by these key parameters is incorporated into the controller.

Ignoring the dependency on Reynold's number essentially fails to account for changes in air density. The effect of changing air density is less marked than changes in incidence and airspeed, and it is not used as a key parameter in the interests of keeping the controller complexity down. There are also other candidate key parameters which are also not used in this design study, examples being aircraft weight and centre of gravity location.

3.5.3 Engine governor

The engine modelled in the GVAM was not developed for automatic control and as such has a highly non-linear throttle demand to achieved thrust characteristic. The non-linear response is primarily due to a governor which becomes active at high engine speeds in order to limit the temperature. Figure 3.3 shows the change in dynamics and gain which occur when the governor becomes active. FNP is engine fan speed, and is

essentially a measure of thrust, and APTHTP is the throttle servo demand.

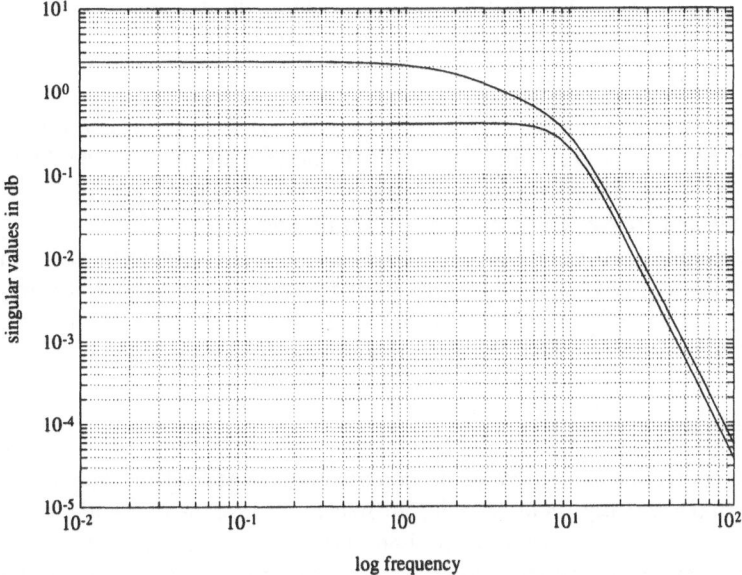

Fig. 3.3. Transfer function form APTHTP to FNP for governor active and inactive.

For a pilot flying the aircraft manually this is not an undesirable characteristic in that the reduced gain of FNP/APTHTP at high engine speeds gives a finer thrust setting ability, which is useful when in the hover. However, this characteristic poses potential problems for a closed-loop controller as it appears as an abrupt change in gain of ×10 at mid and low frequencies.

To counter the effect of the governor an open-loop non-linear engine controller is used to achieve a roughly linear response between demanded and achieved thrust. Figure 3.4 shows the structure on the engine controller.

The engine controller detects if the governor is active or not, and sets the switch S accordingly. $D(s)$ is chosen so that it appears as if the dynamics of the engine do not change when the engine governor cuts out. If $D_H(s)$ and

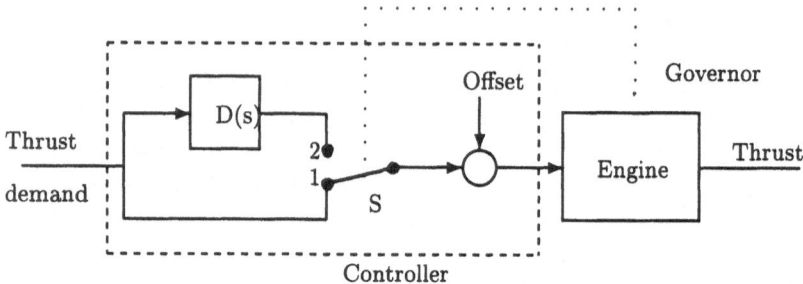

Fig. 3.4. Engine controller

$D_L(s)$ are the dynamics between APTHTP and the fuel flow to the engine for the governor active and not active respectively, then

$$D(s) = D_H(s)D_L^{-1}(s)$$

When the switch S is operated it must be ensured that there is no discontinuity in the throttle demand, APTHTP. To do this an offset on APTHTP is used; every time the governor cuts in or out, the required value of OFFSET is recalculated to give a continuous APTHTP signal.

This engine controller has been found to perform very well, and in particular does not have a tendency to limit cycle if operated at the governor switching point. It enables all the linear design to be carried out without without having to consider the engine governor.

3.5.4 Thrust resolution

A linear controller design could be done using a linearisation which has the plant actuators as inputs i.e. nozzle, throttle and tailplane demands. However this is not a good approach as the linearised model is highly dependent on the nozzle angle; to see this refer back to Figure 3.1. If the nozzle angle, ATHDFP, changes, then clearly the effect of a unit change in thrust, T, on accelerations $\dot{u}$ and $\dot{w}$ is changed; in the extreme cases of ATHDFP=0 and ATHDFP= 90°, then a change in T only effects $\dot{u}$ or $\dot{w}$ respectively. Similarly a change in T alters the effect of a unit change in ATHDFP on $\dot{u}$ and $\dot{w}$. If instead for the purposes of linear design we replace ATHDFP and T with resolved thrust demands in the directions of u and w, then this non-linear effect is avoided. If the engine controller thrust demand is denoted THRSTA, then

AXF = THRSTA cos ATHDFP

AZF = THRSTA sin ATHDFP

replace THRSTA and ATHDFP as plant inputs for the linear design. The implemented control law then simply calculates THRSTA and ATHDFP from AXF and AZF.

3.5.5 Bank angle compensation

With the nozzles down, if the aircraft banks over then the vertical component of the thrust is reduced;

$$T_z = T \cos \Phi$$

where Φ is the bank angle. The effect of this is for the aircraft to lose height, which is particularly undesirable when attempting to manoeuvre to a landing spot in hover mode. To counter this effect, the controller demand to the engine is divided by $\cos \Phi$. This compensation is blended out as the nozzles go aft.

3.5.6 Other non-linearities

There are many other non-linear effects which are harder to take account of in the design. One particularly important effect is the reingestion of engine exhaust gasses as the aircraft approaches the ground in the hover; this causes a rapid loss in thrust which must be countered with an increase in throttle demand. Another aerodynamic effect is the vectored thrust impinging on the tailplane as the nozzles go aft. These type of effects can be viewed as disturbances which the controller must reject.

3.6 Uncertainty modelling

3.6.1 Unstructured uncertainty

The loop-shaping/$\mathcal{H}_\infty$ design procedure introduced in §2.3.4 makes the system robust to additive perturbations to the normalised left coprime factors of the weighted plant. As has already been discussed, this represents an unstructured uncertainty description. A direct physical interpretation of robustness to coprime uncertainty is not obvious, but McFarlane gives bounds on other forms of unstructured uncertainty. Equations 3.1, 3.2 and 3.3 from [21] are bounds on multiplicative output, multiplicative input, and additive robustness measures (defined in §2.3.1) respectively.

$$GK(I + GK)^{-1} \leq 1 + \gamma \overline{\sigma}(\tilde{M}_s)\kappa(W_2) \tag{3.1}$$

$$K(I + GK)^{-1}G \leq \gamma \overline{\sigma}(\tilde{N}_s)\kappa(W_1) \tag{3.2}$$

$$K(I + GK)^{-1} \leq \overline{\sigma}(\tilde{M}_s)\overline{\sigma}(W_1)\overline{\sigma}(W_2) \tag{3.3}$$

$(\tilde{M}_s, \tilde{N}_s)$ is a normalised left coprime factorisation of the shaped plant, W_2GW_1, and $\kappa(\bullet)$ denotes condition number, $\frac{\overline{\sigma}(\bullet)}{\underline{\sigma}(\bullet)}$. Using the following equalities from [21]

$$\overline{\sigma}(\tilde{M}_s) = (\frac{1}{1 + \underline{\sigma}^2(W_2GW_1)})^{1/2} \tag{3.4}$$

$$\bar{\sigma}(\tilde{N}_s) = (\frac{\bar{\sigma}^2(W_2 G W_1)}{1 + \bar{\sigma}^2(W_2 G W_1)})^{1/2} \qquad (3.5)$$

it can be seen that if γ is small and $\kappa(W_1)$ and $\kappa(W_2)$ are made small, then tight bounds on multiplicative input and output robustness are achieved. These three forms of uncertainty have a more direct physical interpretation. For example, multiplicative input robustness overbounds multiplicative uncertainty to each actuator on its own e.g. robustness to multiplicative uncertainty on the first actuator is given by:

$$1/\bar{\sigma}([\ 1 \quad 0 \quad \cdots \quad 0\ ] K(I + GK)^{-1}G \begin{bmatrix} 1 \\ 0 \\ \vdots \\ 0 \end{bmatrix}) \geq 1/\bar{\sigma}(K(I + GK)^{-1}G)$$

Hence careful choice of weighting functions W_1 and W_2 can ensure that the resulting closed-loop system has good robustness to actuator and sensor uncertainty, and also unmodelled plant dynamics. This is discussed in more detail in chapter 4.

3.6.2 State-space uncertainty modelling

Changes in the state space description of the system as key variables change can be used to describe structured system uncertainty. If the nominal system has state space matrices $[A, B, C, D]$, then the perturbed system can be written as:

$$\dot{x} = (A + dA)x + (B + dB)u$$

$$y = (C + dC)x + (D + dD)u$$

This uncertainty can be written as a linear fractional transformation as is shown in [23] with the form of Figure 3.5 where $\delta = 0$ gives the nominal

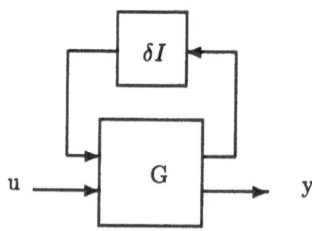

Fig. 3.5. State-space uncertainty as an LFT.

system, and $\delta = 1.0$ gives perturbed system state space matrices. Once in this form, the state-space uncertainty can be used in structured singular value analysis and also for design with D-K iteration. State space variations due to changes in airspeed have been used for a D-K iteration design on the GVAM in [18]. This form of uncertainty modelling is also used here in chapter 6 for the analysis of a gain schedule.

3.7 A specification for linear feedback design

In this section specifications for linear designs on the GVAM are drawn up based on the proceeding sections of this chapter. Arriving at quantitative requirements is not always straightforward, but an attempt to do so is made when possible. Required handling qualities do not necessarily need to be closely adhered to at this stage as pilot demand precompensators can be used to tailor them later. However, if some attention is paid to them during the design of the feedback controller, then it is likely that much simpler demand precompensators will suffice. Hence an attempt is make to meet both robustness and handling quality requirements with the feedback controller if possible.

Inputs

1. <u>Hover and transition.</u> Inputs for the linear models are taken to be resolved thrusts, AXF and AZF, and tailplane demand ETADA.
2. <u>Wingborne flight.</u> Inputs are engine controller demand, THRSTA, and tailplane demand, ETADA.

Outputs

1. <u>Hover and transition.</u> Outputs are ground speed (VHOR), vertical speed (VKD), and pitch attitude (THETD). The choice of ground speed as opposed to airspeed reflects typical manoeuvres in the hover such as positioning relative to a landing pad, and also gives required response to wind gusts at low speed. Choosing VKD is more sensible than flight path angle (γ) because γ depends on forward speed, and so linear models taken for increasing speeds would show increasing gain from this output. Pitch attitude is chosen because it enables pitch to be regulated by the controller leaving the pilot with just a two-inceptor demand system.
2. <u>Wingborne flight.</u> Outputs are forward airspeed (VTKT) and pitch rate (Q). Control of these quantities gives a conventional response to pilot controls whilst in addition attaining speed hold, approximate attitude hold, and decoupling.

Robustness requirements

Whilst meeting performance requirements, ideally robustness should be max-
imised in some sense. In the absence of detailed knowledge of where the key
uncertainty is in the system, a robustness measure which captures a broad
enough class of perturbations should be used. The coprime factor uncertainty
meets this requirement, and is the measure of uncertainty chosen to be minimised for the designs in Part II of this monograph. Note that this is the same
philosophy that it often used in classical SISO design where good gain and
phase margins are aimed at without reference to any detailed uncertainty
modelling. As has already been pointed out, coprime factor uncertainty is
possibly more appropriate than the more conventional additive and multiplicative robustness measures in that it allows the number of unstable poles
of the plant to vary.

      As a check for robustness to actuator and sensor uncertainty, the multi-
plicative input and output robustness should be evaluated. Designs should
have good robustness to changes in candidate key variables which are not
used for gain scheduling or switching. Robustness to key variables which are
used is highly desirable as this reduces the number of design points required
for the whole flight envelope.

Handling Qualities

1. <u>Hover and Transition.</u> To meet the handling qualities criteria in §3.4.2:
   - Ensure that demand type and scaling reflects actuator potential.
   - Keep the lag between demands and actuators to a minimum.
   - Time responses should have little or no overshoot, and exhibit no resid-
     ual oscillations.
   - 90% of commanded attitude should be achieved in 1 to 2 seconds.
2. <u>Wingborne Flight</u> To meet the handling qualities criteria in §3.4.3:
   - Minimise the closed-loop phase lag between demands and controlled
     outputs; include a direct term in the forward path.
   - Pitch channel bandwidth $\equiv \omega_{sp} = \frac{2\pi}{1} = 6 rad/s$.
   - Phase rate at 180 degrees should be kept small, where the 180 degrees
     refers to the transfer function from pitch rate demand to pitch attitude
     output.

Decoupling

Whilst addressing all of the above, any decoupling which can be achieved is
highly desirable.

3.8 Summary

VSTOL aircraft operational requirements have been reviewed, and the Generic VSTOL Aircraft Model introduced as a case study for the design approaches developed in part I. The chief non-linearities of the GVAM have been outlined, and the action taken to account for them discussed. Related to this is plant uncertainty modelling, which combined with a review of handling quality requirements has been used to draw up a specification for the linear designs. This specification is used in the next chapter for the loop-shaping design examples on the GVAM.

CHAPTER 4
ROBUST DESIGN USING LOOP-SHAPING

4.1 Introduction

Loop-shaping is well accepted and widely used for scalar feedback design. For a scalar system it is a straightforward task to convert closed-loop disturbance rejection requirements into loop-shape requirements. However, when the approach is generalised for multivariable systems this task is somewhat more complicated. The problem arises in that different loops may have very different gains which in turn means that the directionality of references, disturbances and plant uncertainty become very important when specifying and evaluating stability and performance. An effect of different loop gains is that a compensator used to give good properties at one break point in the loop may give poor properties at another. To see this, consider Figure 4.1 where a compensator K is to be designed by loop-shaping to give good rejection of disturbances d_1 and d_2. Consider shaping GK i.e. we break the loop at point

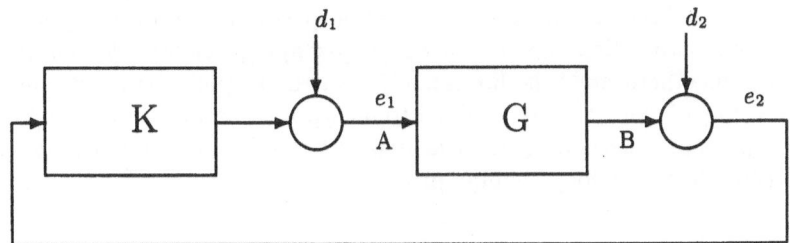

Fig. 4.1. Loop breaking points

B. If the condition number of G, defined as

$$\kappa(G) \triangleq \frac{\bar{\sigma}(G)}{\underline{\sigma}(G)} \tag{4.1}$$

is large with $\underline{\sigma}(G) << 1$ and $\bar{\sigma}(G) >> 1$ over the frequency range for which disturbances are to be rejected, then designing K to increase gain in the directions for which the gain of G is small so that $\underline{\sigma}(GK) \gg 1$ for this frequency range will give good rejection of d_2 disturbances. If we now consider breaking the loop at point A, it can be seen that the rejection of d_1 is dictated by the gain of KG. If K has been designed only to increase gain in certain directions, KG may have low gain in certain directions i.e. $\underline{\sigma}(KG) \ll 1$ over the frequency range for which disturbances are to be rejected. Hence rejection of disturbances at A may be much poorer than rejection of disturbances at B. In the case that $\kappa(G) \simeq 1$ and K is chosen such that $\kappa(K) \simeq 1$, then $\kappa(GK) \simeq \kappa(KG) \simeq 1$, and properties at one loop breaking point will reflect those at another.

In §4.2 a procedure for selecting loop-shaping weights for the methodology of McFarlane and Glover described in chapter 2 is outlined. This procedure has been developed from experience with loop-shaping designs for the Generic VSTOL Aircraft Model (GVAM). Two example designs on the GVAM are presented in sections 4.6 and 4.7 to illustrate the procedure. The first one is for a forward speed of 6 knots, and represents the hover/powered-lift mode. The second is for fully wing-borne flight, and as such presents a very different design problem. For both of these examples the plant is well conditioned at the desired bandwidth. To illustrate the procedure applied to a poorly conditioned plant, an example design on a fractional distillation column is given in §4.8.

Freudenberg [5] has examined the problem of multivariable loop-shaping for ill-conditioned plants. His approach is to examine the dependence of closed-loop properties at one point in the loop on the open-loop singular values at another. This leads to conditions which must be satisfied if properties at both points are to be good, and hence shows how to shape the singular values at one break point to ensure good properties at both break points. In §4.3 some justification is given for the loop-shaping weights selection procedure outlined here, and it is shown how it relates to Freudenberg's conclusions in §4.4. In §4.5 methods are outlined to incorporate extra measurements into the loop-shaping design. Extra measurements result in non-square plants and thus complicate the loop-shaping procedure.

4.2 A loop-shaping procedure

4.2.1 The procedure

The following procedure has been developed with experience on designs on the GVAM. It has been found to give controllers with good robust performance. This makes them particularly suitable for implementation in that fewer controllers are needed to meet performance specifications across the whole flight envelope. Step 5 refers to the "align" algorithm. The "align" algorithm, originally proposed in [38], approximately inverts the plant at a chosen frequency [38] using a real rational matrix A_L. Suppose the plant $G(s)$ is to be aligned at frequency ω_1. Then we require

$$G(j\omega_1)A_L \simeq I$$

Let A_L be described by its column vectors:

$$A_L = (a_1, a_2, \ldots, a_m)$$

and $G(j\omega_1)$ by its row vectors:

$$G^T(j\omega_1) = (v_1, v_2, \ldots, v_m)$$

In order to invert $G(j\omega_1)$ we need $v_j^H a_i = 0$ for $i \neq j$ where x^H denotes conjugate transpose of x. Maximising the following expression for each column of A_L attempts to achieve this, and is known as the "align algorithm":

$$\max_{a_i} \frac{|v_i^H a_i|^2}{\Sigma_{j \neq i} |v_i^H a_i|^2}$$

This can be solved as a generalised eigenvalue problem[9], and is a function available in the Matlab Multivariable Frequency Domain Toolbox [39].

1. Scale all outputs such that one unit of cross-coupling into each of the outputs is equally undesirable.

2. Scale all inputs to reflect the relative actuator bandwidth capabilities. This step dictates the relative use of each of the actuators if the "align" in step 5 is not used. The scaling is also very important for the interpretation of the multiplicative input robustness in step 7. The relative actuator capabilities are usually dictated by the speed of their nominal dynamics. However, sometimes uncertainty may restrict the bandwidth at which an actuator can be used. Another consideration is that if an actuator is prone to saturation this might dictate the bandwidth up to which it is used. The scalings may be only rough for the preliminary design. Once actuator use has been checked from time and frequency responses, the scalings can be altered in a very intuitive manner to change the relative actuator usages.

3. Re-order the inputs so that the plant is as diagonal as possible. This makes designing the pre-compensator easier in that a diagonal weighting may suffice.

4. Alter the singular value roll-off rates at the desired cross-over frequency. To do this plot the transfer functions :

   a) from each actuator alone to all outputs, and
   b) from all actuators to each output alone.

   By examining these transfer functions select the elements of the diagonal pre- and post-compensators W_P and W_2 such that the singular values of $W_2 G W_P$ have a roll-off of approximately 20dB/decade at the desired bandwidth(s).

   Add integral action to the terms of W_P, and high-frequency roll-off terms if required.

5. Align the singular values at the desired bandwidth. This optional step is only carried out if the plant is well-conditioned (see §4.2.2). The MATLAB Multivariable Frequency-Domain Toolbox *align* [39] algorithm can be used to find the align matrix, W_A. The shaped plant is thus as in Figure 4.2.

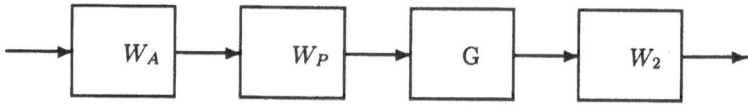

Fig. 4.2. The pre- and post-compensated plant.

6. Robustness optimisation : perform a left-coprime factor stabilisation on $G_S = W_2 G W_1$ where $W_1 = W_P W_A$. Check the optimal gamma, γ_{min}. This is used as a design indicator; a γ_{min} less that about 4 is usually acceptable, and a higher γ_{min} indicates that the specified singular values are inconsistent with robust stability requirements. Note that a large γ_{min} does not necessarily indicate a poorly conditioned plant.
   If some structure is known about plant uncertainty then a DK-iteration (see §2.4) can be carried out here as described in [18].
7. Form the closed-loop, and check the time responses for rise-times, interaction levels, and actuator use. Check the robustness of the design with respect to multiplicative input and output robustness. Care must be taken when evaluating these measures to scale the plant correctly; for example when evaluating input robustness the actuators should be scaled to reflect their relative bandwidths. If they are scaled in widely different units, then an apparently high gain channel may cause robustness problems by feeding onto an apparently low gain channel.
   Robust performance may also be evaluated by performing step responses with off-design point plants.
8. If required, alter W_1 and W_2 and re-run the optimisation to :
   a) reduce/increase the use of a particular actuator
   b) alter the rise-times,
   c) alter the relative levels of interaction between outputs.

4.2.2 Ill-conditioned plants

As discussed in §4.1, aligning the singular values at one point in the loop may lead to poor robustness and performance properties at another point. In particular, an ill-conditioned plant will give rise to an align matrix with high condition number, and lead to potentially poor robustness. Following the above procedure, but missing out stage 5 has been found to produce designs with comparable performance with other design methods for an ill-conditioned design example studied (see §4.8). With ill-conditioned plants it may still be possible to align a subset of the plant inputs and outputs.

4.2.3 Controller positioning in the loop

So far only the open-loop singular values have been considered, and the feeding in of references into the loop has not been discussed. The conventional way of entering references into the loop is shown in Figure 4.3. For $\mathcal{H}_\infty$ one

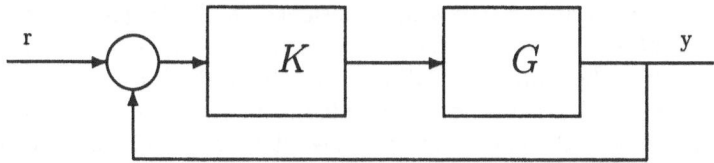

Fig. 4.3. Conventional implementation of K

degree of freedom design formulations this is normally the only place that K_∞ can be placed if it has integral action. However, with the loop-shaping approach there is more flexibility in that the K_∞ and the loop weighting functions can be kept separate. Figure 4.4 shows the conventional set-up for a loop-shaping design, and Figure 4.5 an alternative. Notice that in the

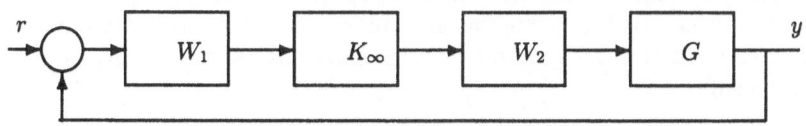

Fig. 4.4. Conventional implementation for a loop-shaping controller

set-up in figure 4.5 integral action would all be placed in the W_1 weighting function, and so the steady-state error will be zero as required. Bounds given by McFarlane in [21] show that K_∞ will be well-conditioned, so multiplying the references by $K_\infty(0)$ does not cause numerical problems.

The observer implementation of loop-shaping design controllers outlined in chapter 6 feeds the references in at yet another alternative place in the loop. It has been found with all the design studies carried out so far that the structure shown in Figure 4.5 and the observer approach give the best responses, primarily because they are not prone to produce overshoot to step

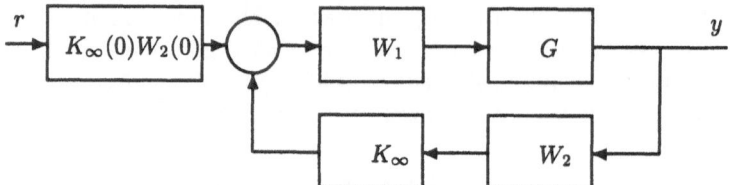

Fig. 4.5. Alternative implementation for loop-shaping controllers.

responses. The conventional set-up in figure 4.4 has been found often to give large amounts of overshoot. This is because the references directly excite the dynamics of K_∞, and K_∞ has been designed in the frequency domain for robustness objectives rather than time-domain performance. The set-up in Figure 4.5 has the same motivation as in classical PID control where the derivative feedback term is not directly fed from the error signal. It is shown in [40] that one of the two observer forms (there are two: one is the dual of the other) gives a closed-loop of $\tilde{N}$ where $\tilde{N}$ is one of the coprime factors of the plant. The good time-domain properties follow from $\tilde{N}$ being normalised which means that it is close to unity at low frequency, and that it does not peak at high frequency.

4.2.4 Scaling

As is alluded to in stage 2 of the design procedure, careful scaling of the linearised plant is of great importance. Scaling is not just important for specification of relative actuator use and output cross-coupling trade-offs, but also for analysis of a design in the frequency domain. As an example, consider the multiplicative input robustness measure $1/\overline{\sigma}[K(I + GK)^{-1}G]$. As discussed in §3.6, this can be interpreted as an overbound on robustness to multiplicative uncertainty on the actuator transfer functions i.e. the perturbed plant is of the form

$$ G \begin{bmatrix} 1 + \delta_1 & & & \\ & \ddots & & \\ & & \ddots & \\ & & & 1 + \delta_u \end{bmatrix} $$

where the δ_i's are complex scalars. The unstructured input robustness measure allows for off-diagonal elements as well as the diagonal δ's above. If the inputs are scaled in very different units, then this implies that the off-diagonal δ's can allow large signals to couple into small signals at the plant input; this will make the off-diagonal δ's dominate the unstructured robustness test, and give very conservative results as to the systems robustness to true uncertainty.

4.3 Justification of the procedure

The outlined procedure has proven straightforward to use, and quickly gives a good idea of what performance is achievable with a particular plant. The fact that no γ iteration is required in the optimisation step is advantageous in that design iteration with different loop weighting functions is quick to perform. This is unique to the normalised coprime factor robust stabilisation problem formulation, and other $\mathcal{H}_\infty$ techniques require iteration on γ. In practice few, if any, iterations on the weights are required; for example the design in §4.8 for an ill-conditioned plant is the first design produced, and this is comparable in performance to a design in [41] making use of the structured singular value. The selection of weights is straightforward, and uses familiar SISO loop-shaping guidelines. For example, ensuring a cross-over roll-off close to 20dB/decade is consistent with Bode's observation [7] that roll-off rate determines phase, and that a rate of 20dB/decade corresponds to 90 degrees phase. Hence this should ensure good phase margins. A discussion of generalising Bode's gain/phase constraints to the multi-input multi-output case can be found in [42].

In example designs the procedure has given good robust stability and performance. The robust stability follows from ensuring a low γ; a large γ indicates that the specified loop-shapes are incompatible with robust stability. It is shown in [16],[17] that the controller does not cancel plant poles which is consistent with the desire for robust performance. Some other $\mathcal{H}_\infty$ formulations do cancel plant poles, and this is a particular problem if the plant has lightly damped poles near the axis as is shown in [10] for a VSTOL aircraft and in [43] for a F/A-18 flight controller design study.

The robust performance property follows in some loose sense from the fact that the $\mathcal{H}_\infty$ optimisation can be regarded as rejecting disturbances entering the loop at two different points; the normalised left coprime-factor robust stabilisation problem is to minimise

$$\epsilon_{\max}^{-1} = \gamma_{\min} = \inf_K \left\| \begin{bmatrix} K \\ I \end{bmatrix} (I - GK)^{-1} \tilde{M}^{-1} \right\|_\infty \tag{4.2}$$

McFarlane points out [20] that we can right-multiply this by the inner function $[\tilde{N}, \tilde{M}]$ without altering the $\mathcal{H}_\infty$ norm to get

$$\epsilon_{\max}^{-1} = \gamma_{\min} = \inf_K \left\| \begin{bmatrix} K \\ I \end{bmatrix} (I - GK)^{-1} [I, G] \right\|_\infty \tag{4.3}$$

This is often referred to as an $\mathcal{H}_\infty$ four-block problem, and corresponds to the closed-loop transfer function from $\begin{bmatrix} d_2 \\ d_1 \end{bmatrix}$ to $\begin{bmatrix} e_1 \\ e_2 \end{bmatrix}$ in Figure 4.6 with $\triangle_1 = \triangle_2 = 0$.

Freudenberg shows in [44] that robust performance is synonymous to tolerating simultaneous uncertainty at different points in the loop. Figure 4.6

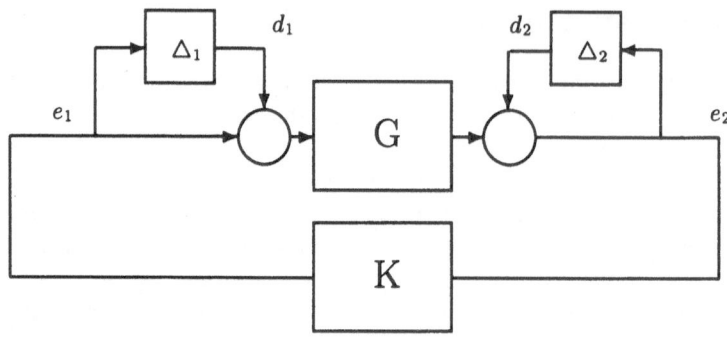

Fig. 4.6. Simultaneous uncertainty at two points in the loop.

illustrates simultaneous uncertainty at two different points in the loop. A solution to the problem of maximising robustness to simultaneous Δ_1 and Δ_2 can be found using the structured singular value. However the above $\mathcal{H}_\infty$ four-block problem formulation gives a conservative solution to this problem, and hence some degree of robust performance could be expected from a four-block design. To see intuitively why this is, consider for example the $(I - GK)^{-1}$ term as corresponding to d_2 disturbance rejection, and the $K(I - GK)^{-1}G$ term as corresponding to input multiplicative robustness. Then we can say that output disturbance rejection should remain good in the face of input multiplicative uncertainty i.e. it shows robust performance.

Of course it must be remembered that these disturbances are on the shaped plant, not the actual plant. However, the design procedure given uses essentially round weights i.e. weights of small condition number, and hence good disturbance rejection at the shaped plant inputs and outputs indicates good disturbance rejection at the actual plant inputs and outputs. Bounds are given in [21] on the singular values of the controller K_∞ from the normalised coprime factor robust stabilisation procedure which show that the controller is well-conditioned provided that the achieved γ is low. Hence $\kappa(K) = \kappa(W_1 K_\infty W_2)$ will be small if the design procedure given in §4.2 is followed because $\kappa(W_1)$ and $\kappa(W_2)$ will be small.

When applying the procedure to ill-conditioned plants, the fact that the loop weights chosen are round reflects the plant directionality, and produces realistic designs given the plant's inherent limitations. This approach is consistent with Freudenberg's conclusions in [5]. In the next section his results are reviewed, and then related to the loop-shaping procedure given here.

4.4 Freudenberg's guidelines

The desirability of round controllers and plants is easily seen from [5, Proposition 4.1] :

Proposition 4.4.1 (Freudenberg 1990). *Assume that det[G] $\neq$ 0 and det[K] $\neq$ 0, then each singular value of the input sensitivity function satisfies the bounds*

$$(1/\kappa[G])\sigma_i[S_O] \leq \sigma_i[S_I] \leq \kappa[G]\sigma_i[S_O] \tag{4.4}$$

$$(1/\kappa[K])\sigma_i[S_O] \leq \sigma_i[S_I] \leq \kappa[K]\sigma_i[S_O] \tag{4.5}$$

If $\kappa(K) = 1$, then $\sigma_i[S_O] = \sigma_i[S_I]$. Hence a rounded compensator ensures similar levels of disturbance rejection at the plant input and output. If $\kappa(K) >> 1$, then the properties at the input may be arbitrarily poor when the properties at the output are good.

Freudenberg points out that we must consider robustness as well when designing the loop shapes. For example $K = G^{-1}(s)$ gives identical nominal properties at the input and output, but it is well known that such a controller is likely to show poor robustness properties. To address this he considers singular value decomposition of the plant into high and low gain subsystems

$$G = X_1 T_1 Z_1^H + X_2 T_2 Z_2^H \tag{4.6}$$

where $T_1 = diag[\tau_1 \ldots \tau_l]$, $T_2 = diag[\tau_{l+1} \ldots \tau_n]$, $X_1, Z_1 \in C^{n\times l}$, $X_2, Z_2 \in C^{n\times(n-l)}$ and it is assumed that $\underline{\sigma}[T_1] >> \overline{\sigma}[T_2]$. It is also assumed for simplicity that $\overline{\sigma}[T_i] = \underline{\sigma}[T_i]$, $i = 1, 2$ i.e. each subsystem has a uniform gain. Let the open-loop singular values of GK be partitioned conformally with the partition of G as:

$$L_O = GK = V_1 \Sigma_1 U_1^H + V_2 \Sigma_2 U_2^H \tag{4.7}$$

where $V_1, U_1 \in C^{n\times l}$, $V_2, U_2 \in C^{n\times(n-l)}$, $\Sigma_1 = diag(\sigma_1 \ldots \sigma_l)$, and $\Sigma_2 = diag(\sigma_{l+1} \ldots \sigma_n)$. It is then shown that if the output sensitivity is to be shaped to prevent the sensitivity at the plant input from being large, then we must have

$$\overline{\sigma}(X_2^H S_O X_1) = \overline{\sigma}(X_2^H T_O X_1) << 1 \tag{4.8}$$

It is also shown that if the input sensitivity is to be robust to output multiplicative perturbations then the following condition must also be satisfied:

$$\overline{\sigma}(X_2^H T_O)\overline{\sigma}(S_O W_1) << 1 \tag{4.9}$$

Freudenberg then considers the relationship between open-loop properties at one point in the loop and closed-loop properties at another. This leads to the conclusion that if equation 4.9 is to be satisfied, then the left singular subspaces of L_O and G must satisfy

$$V_i = X_i, i = 1, 2 \tag{4.10}$$

and also

$$\sigma_l(L_o) >> 1 >> \sigma_{l+1}(L_o) \tag{4.11}$$

must be satisfied over some frequency range. Hence the rejection of disturbances in some directions will not be good, but this can be seen as a result of the plant ill-conditioning. The designer is thus free to choose the right singular subspaces of L_o. The rejection of disturbances in the subspace $\mathbf{X_2}$ is limited, and so we could try to choose the right singular subspaces such that

$$\mathbf{Y_d} \subset \mathbf{U_1} \qquad (4.12)$$

i.e. the disturbances feed through into the high gain subsystem. A unitary matrix can be used as a precompensator to alter the right singular subspaces; this will satisfy the partition in equation 4.11, and $\mathbf{V_i} = \mathbf{X_i}$ is automatically satisfied.

This ties in well with the loop-shaping procedure given here in that round compensators are favoured. The last stage of altering the right singular subspaces has no effect on the normalised robust stabilisation design if it is done using a constant unitary matrix i.e. a rotation matrix. To see this consider postmultiplying G with a constant unitary matrix P_R. Let $G = \tilde{M}^{-1}\tilde{N}$ where $(\tilde{M}, \tilde{N})$ is a normalised left coprime factorisation of G. Then a normalised left coprime factorisation of GP_R is given by $(\tilde{M}, \tilde{N}P_R)$. Hence the LCF optimisation becomes

$$\epsilon_{\max}^{-1} = \gamma_{\min} = \inf_K \left\| \begin{bmatrix} K \\ I \end{bmatrix} (I - GP_RK)^{-1}\tilde{M}^{-1} \right\|_\infty \qquad (4.13)$$

If this is then right multiplied by the unitary matrix

$$\begin{bmatrix} P_R & 0 \\ 0 & I \end{bmatrix} \qquad (4.14)$$

the optimisation becomes to find

$$\epsilon_{\max}^{-1} = \gamma_{\min} = \inf_K \left\| \begin{bmatrix} P_RK \\ I \end{bmatrix} (I - GP_RK)^{-1}\tilde{M}^{-1} \right\|_\infty \qquad (4.15)$$

Hence postmultiplying G with P_R has no effect on the final implemented controller.

4.5 Extra measurements

4.5.1 An example system

Often there are additional measurements available to the designer (i.e the plant is tall). For example for the GVAM an additional measurement is the pitch rate, QD. This signal has better signal to noise ratios at high frequency than pitch attitude signal, THETD, and hence should not be discarded.

It is not immediately clear in the loop-shaping context as to how extra measurements should be incorporated into the design. In this section two approaches are examined and it is shown that they both do similar things. The first approach makes use of what are known as complementary filters to combine a position and a rate signal to give a single position signal. This signal is then used for the loop-shaping procedure The second approach is a loop-shaping approach for non-square plants. The two approaches are illustrated and compared on the system shown in Figure 4.7. It consists of a simple mass rotating on a shaft with torque T acting on it, and measurements of angular position, θ and angular velocity, Q, taken.

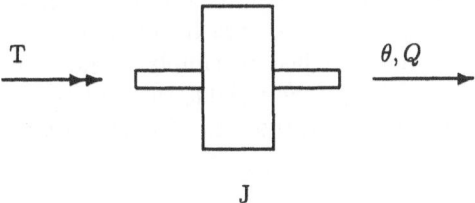

T θ, Q

J

Fig. 4.7. Rotating mass on a shaft acted on by torque T.

The transfer function is thus

$$\theta(s) = \frac{1}{Js^2}T(s) \tag{4.16}$$

4.5.2 Complementary filter approach

Complementary filters are an accepted way of combining two or more signals to be used in different frequency ranges - see for example [45]. A general complementary filter for two signals is of the form

$$\hat{X}(s) = \frac{k}{s+k}X_1(s) + \frac{s}{s+k}X_2(s) \tag{4.17}$$

If $X_1(s) = \theta$ and $X_2(s) = \frac{1}{s}Q$ then we have

$$\hat{X}(s) = \frac{k}{s+k}\theta + \frac{s}{s+k}\frac{1}{s}Q = \frac{k\theta + Q}{s+k} \tag{4.18}$$

If we now perform a loop-shaping design on the plant $G = \frac{1}{s^2}$ (i.e. $W_1 = W_2 = 1$ and $J = 1$) to get a compensator C(s), then we have that the overall feedback is

$$\frac{C(s)(k\theta + Q)}{s+k} = C_1(s)\theta + C_2(s)Q \tag{4.19}$$

where

$$C_1(s) = \frac{C(s)k}{s+k} \tag{4.20}$$

and

$$C_2(s) = \frac{C(s)}{s+k} \tag{4.21}$$

4.5.3 A loop-shaping approach

Consider the plant as

$$G = \begin{bmatrix} \frac{1}{Js^2} \\ \frac{1}{Js} \end{bmatrix} \qquad (4.22)$$

which is the transfer function from the input torque to angular position, and angular rate. A simple loop-shaping approach would be to shape $G_{11} = \frac{1}{Js^2}$ with a precompensator and then just to simply weight $G_{22} = \frac{1}{Js}$ with a simple scalar weight,w; the larger the weight, the more the rate signal gets used. Hence, putting putting $J = 1$, the shaped plant becomes

$$G_s = W_2 \begin{bmatrix} \frac{1}{s^2} \\ \frac{w}{s} \end{bmatrix} W_1 \qquad (4.23)$$

Here W_1 and W_2 are chosen as $W_1 = 1$ and $W_2 = I_2$. The final controller could be split into two components to give

$$u = C_3(s)\theta + C_4(s)Q \qquad (4.24)$$

4.5.4 How the approaches compare

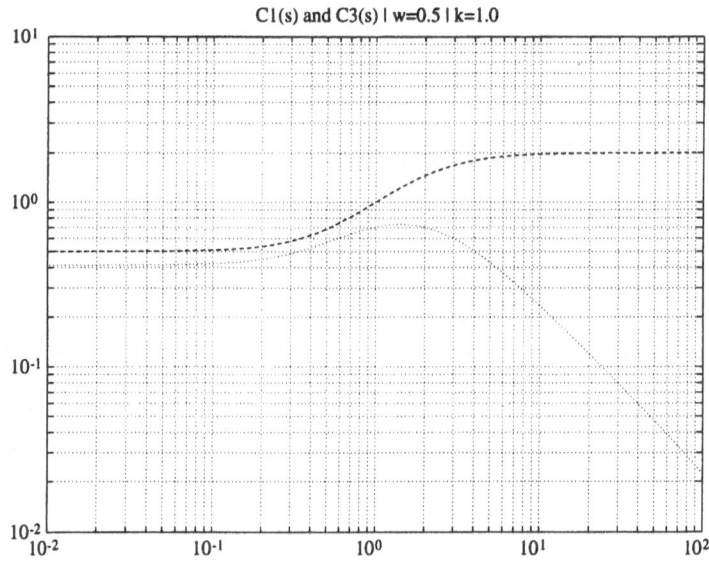

Fig. 4.8. Use of the θ measurement : $C_1(s)$ .... $C_3(S)$ - - -

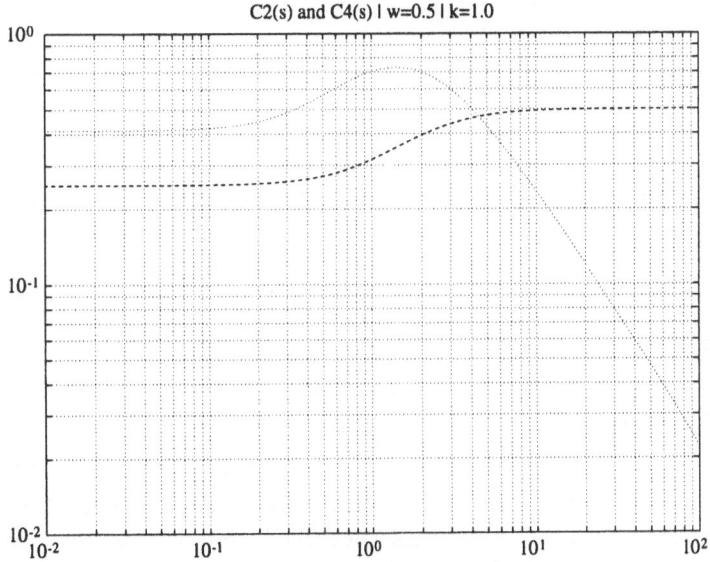

Fig. 4.9. Use of the Q measurement : $C_2(s)$ .... $C_4(S)$ - - -

Figures 4.8 and 4.9 show the angle feedback terms and angular rate feedback terms respectively for the two approaches. For these designs, k was chosen to be 1.0 as this corresponds to emphasising the use of the rate measurement around the bandwidth and higher frequencies (i.e. where the position measurement is less reliable). Correspondingly w was chosen to be 0.5 so as just not to alter the nominal performance-specifying loop shape (i.e. the transfer function from T to θ) around the cross-over frequency.

It can be seen that the angular position measurement is used in a similar fashion up to around cross-over (1 rad/s) for both designs. The loop-shaping design does not roll-off at higher frequencies as the optimal solution has been taken. The complementary filter approach makes more use of the rate signal, but not significantly more. Making k larger would reduce the difference.

Hence it can be seen that when the extra measurement corresponds to the rate of a position measurement, then these two procedures can produce similar controllers. Both approaches have been used for GVAM designs equally successfully.

4.6 GVAM hover design example

4.6.1 Application of the design procedure

This design example is for a linearisation of the GVAM at 6 knots forward speed i.e. for the hover regime. All of the other linear designs for the VSTOL flight mode used in the switched and scheduled control laws developed in the next chapters use similar weighting matrices to those given here for the 6 knot design. Following the design specification arrived at in §3.7, the inputs and outputs are:

Inputs

- AXF - horizontally resolved thrust demand
- AZF - vertically resolved thrust demand
- ETADA - tailplane/reaction jet demand (degrees, range -12 to 12)

AXF and AZF do not have any physical units as they are derived from what is effectively a resolved throttle servo demand. However they have authority limits of approximately 1 unit (the exact authority limits depend on the instantaneous values of AXF and AZF).

Outputs

- VHOR - forward ground speed (feet per second)
- VKD - vertical speed (feet per second)
- THETD - pitch attitude (degrees)

The most important task in this design is selection of the bandwidth. The bandwidth must be high enough to meet the desired performance, but is upper limited by the accuracy of the actuator and sensor models, and their speed of response capacity. The approach taken here is to make the bandwidth just sufficient to meet the level 1 handling quality requirements; in the pitch channel the dominant specification is to reach 90% demanded attitude change in 1 to 2 seconds. In the absence of a corresponding specification for the horizontal and vertical motion cues, this same requirements is applied to them too. The related requirement on the time to reach 63% thrust is not compromised by this. This rise-time requirement translates into a desired bandwidth of $4rads^{-1}$ for all loops. The design procedure is now followed step by step:

1. Output scaling. Using units of feet/sec and degrees achieves the objective of making one unit of cross-coupling equally undesirable for all outputs.
2. Input scaling. Typical variations of ETADA are approximately ten times larger than typical variations on AXF and AZF, so ETADA is scaled by times 10. This scaling is particularly important when evaluating input robustness.

3. Re-ordering plant inputs and outputs. By ordering the outputs VHOR, VKD, THETD and the inputs AXF, AZF, ETADA the plant is made as diagonally dominant as possible at the desired bandwidth.

4. Singular value roll-off rates. By examining Figure 4.10 it can be seen that for the roll-offs to be significantly less than 40dB at the bandwidth requires only phase advance on the ETADA input.

Integral action is added to each channel with zeros at s=3 so as not to alter the roll-off rate at cross-over too much. Thus the overall precompensator is of the form

$$W_p = \begin{bmatrix} \frac{s+3}{s} & 0 & 0 \\ 0 & \frac{s+3}{s} & 0 \\ 0 & 0 & \frac{(s+3)(s+2)}{s(s+20)} \end{bmatrix}$$

Figure 4.11 shows the shaped-singular values so far.

Note that the singular values are already pretty much aligned. This reflects that we have a well-conditioned plant which we have scaled sensibly in steps 1 and 2.

5. Align at the bandwidth. Figure 4.12 shows the singular values after aligning at 4 radians/sec.

6. Optimisation. The normalised coprime factor robustness optimisation gave $\gamma_{\min} = 3.10$. This low value of $\gamma_{\min}$ indicates that our specified loop shapes are consistent with robustness requirements. In this design we choose γ to be 3.5. Choosing γ to be slightly suboptimal can give a great improvement in H_2 performance is shown in [46], and also avoids introducing a fast pole into the controller. An alternative would be to take the optimal controller for which the fast pole goes to infinity, and is therefore replaced by a feedthrough term.

7. Figures 4.13, 4.14, and 4.15 show closed-loop step responses.

These easily meet the specifications in terms of rise-times, overshoot and decoupling levels. Figure 4.16 show multiplicative input and output robustness (see §3.6.1).

Without the plant input scaling on ETADA carried out in step 2, the multiplicative input robustness falls to -17dB. This clearly illustrates the importance of careful scaling if the robustness test is to reflect realistic perturbations to the plant. The poor robustness value which arises here without scaling is due to coupling between loops which appear to have widely different gains whereas in reality they do not.

4.6.2 Reflection of HQ requirements in the procedure

In the handling qualities review in §3.4 it is concluded that handling qualities can be met by:

1. Suitable demand selection and scaling.
2. Keeping lags between references and outputs small.

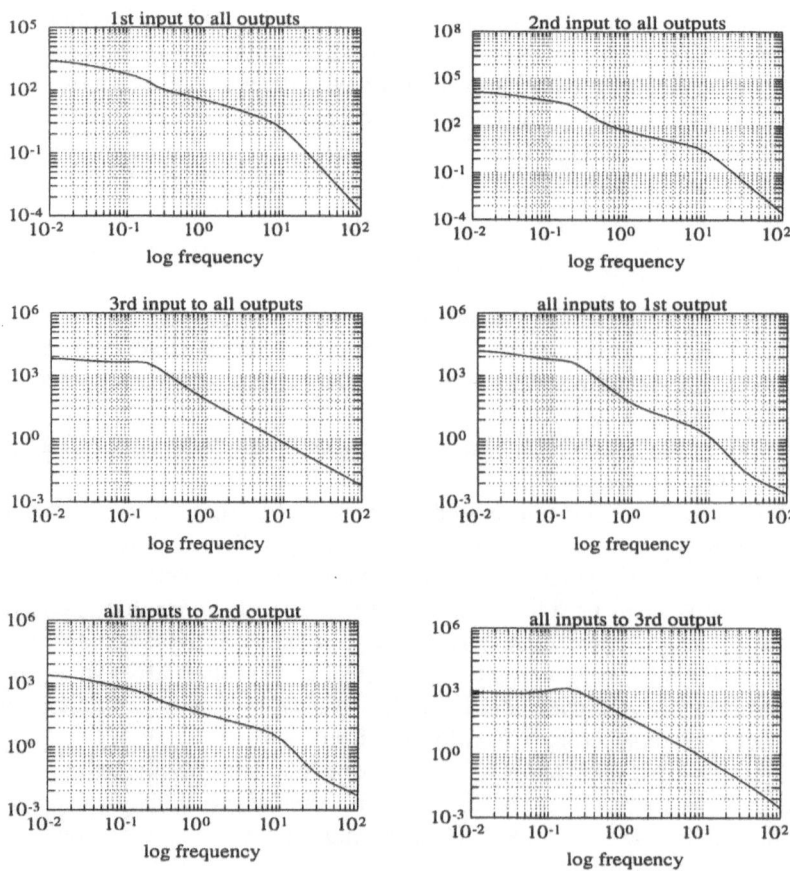

Fig. 4.10. Maximum singular value plots for the nominal 6 knot plant.

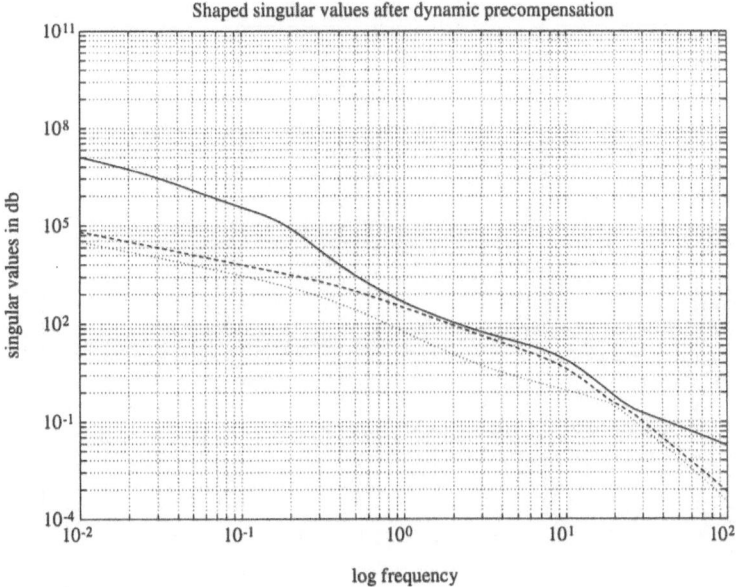

Fig. 4.11. Shaped singular values after dynamic weighting.

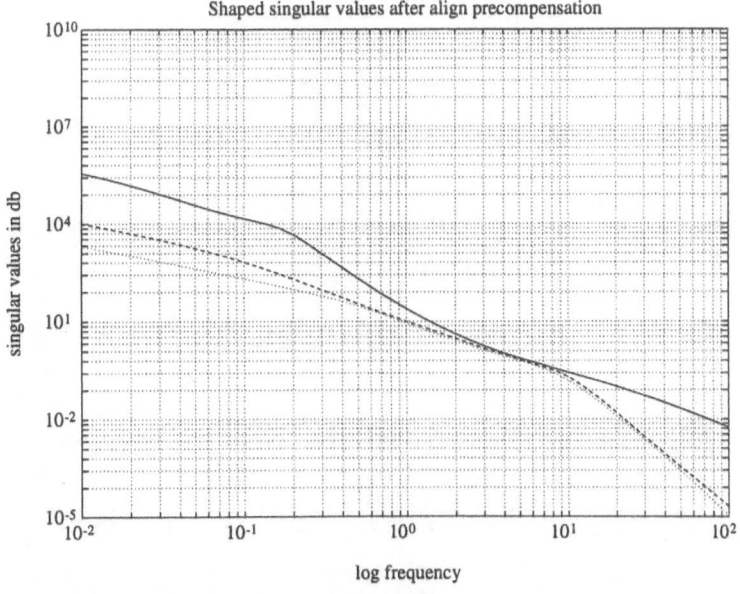

Fig. 4.12. Shaped singular values after aligning.

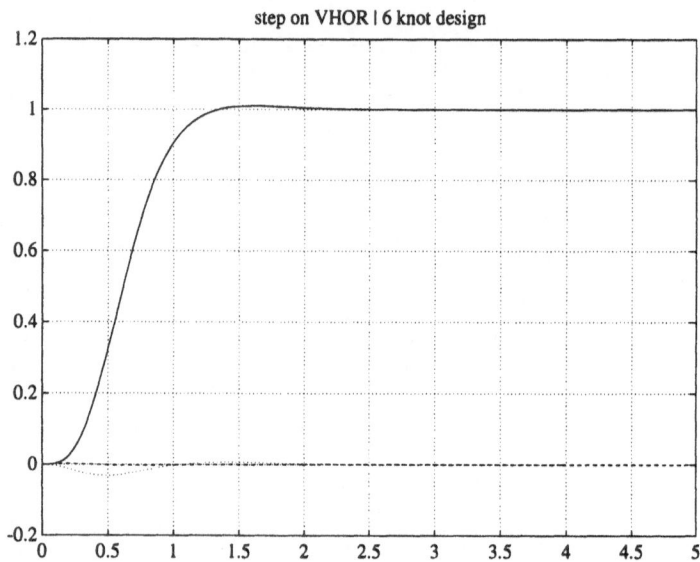

Fig. 4.13. 6 knot loop-shaping design – step on VHOR.

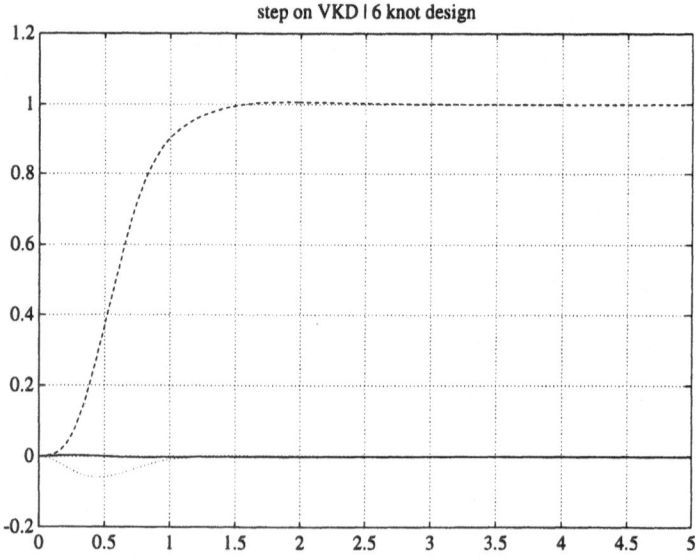

Fig. 4.14. 6 knot loop-shaping design – step on VKD.

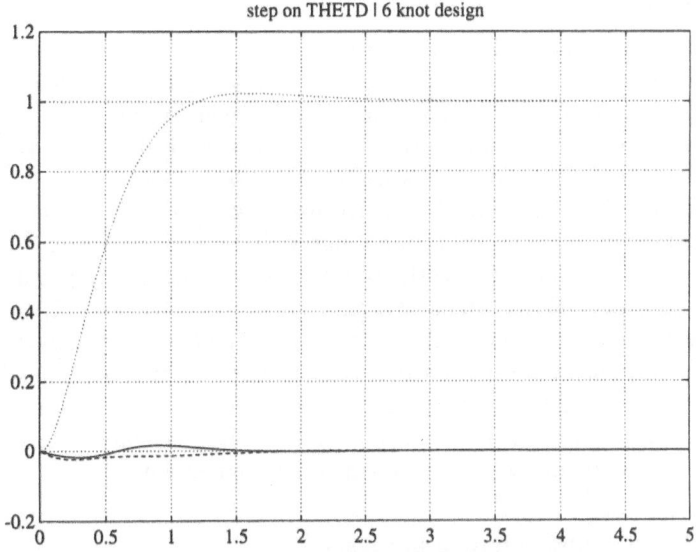

Fig. 4.15. 6 knot loop-shaping design – step on THETD.

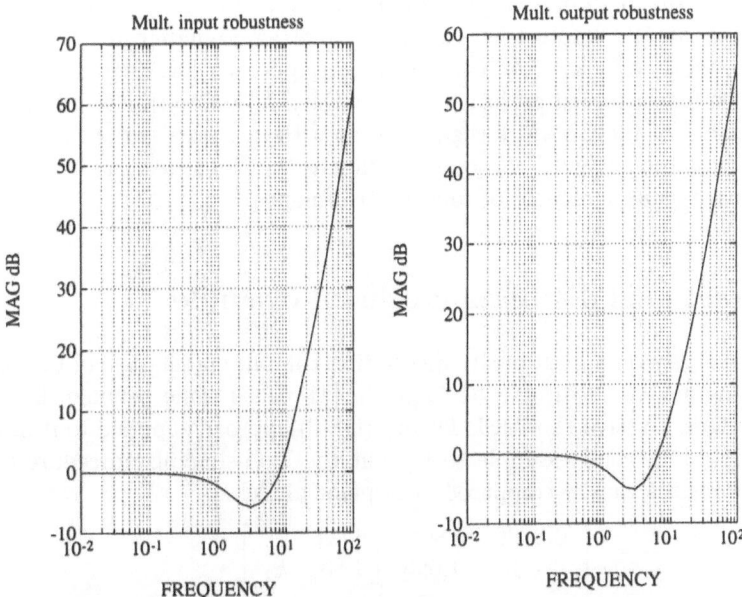

Fig. 4.16. 6 knot loop-shaping design – multiplicative input and output robustness.

3. Achieving time responses with little or no overshoot, and sufficient speed of response.

The first requirement does not impact on the closed-loop controller design, and is therefore not addressed here. The loop-shaping procedure is ideally suited to account for the second two. The designer has direct control of the lag between references and outputs with the use of W_1. This W_1 can be chosen to have the required phase properties, and not to have undesirable poorly damped modes. The K_∞ controller is arrived at from frequency domain robustness considerations, and may well have modes which would give rise to a poor response if directly excited with references; by placing K_∞ in the feedback path this problem is avoided.

Note that the comment in [36] that W_1 should have a direct term to make the response time to pilot demands consistent with the manually flown aircraft is complied with here. Also the designer has direct control over the phase rate at high frequencies via W_1, and so potential pilot induced oscillation problems are easily preventable at the design stage.

4.6.3 Robust performance analysis

For the GVAM it is vital that controllers exhibit a good degree of robust performance since fewer designs will be needed to cover the whole flight envelope. Figure 4.17 shows steps on forward ground speed at 6, 86, and 122 knots.

Clearly the 6 knot design exhibits a good amount of robust performance, as at 122 knots there is appreciable aerodynamic lift. The loop-shaping/coprime factor robust stabilisation approach has been observed to give much better robust performance than the S & KS approach [18]. Robust performance of the GVAM designs is dealt with in more detail in chapters 5 and 6 where the achieved $\mathcal{H}_\infty$ cost function as the flight envelope is traversed is also used as an indicator of robust performance.

4.7 Fractional distillation column example

A design study on a high-purity distillation column making use of the structured singular value, μ, can be found in [41]. This same problem has also had the Quantitative Feedback Theory (QFT) approach applied to it in [47]. Hence this is a good example to try out the loop-shaping procedure on in that the results can be compared. The plant model is

$$G(s) = \frac{e^{-\tau s}}{1 + 75s} \begin{bmatrix} 0.8780 & 0.8640 \\ 1.082 & 1.096 \end{bmatrix} \begin{bmatrix} k_1 & 0 \\ 0 & k_2 \end{bmatrix}$$

where

$$0.8 \leq k_1 \leq 1.2$$

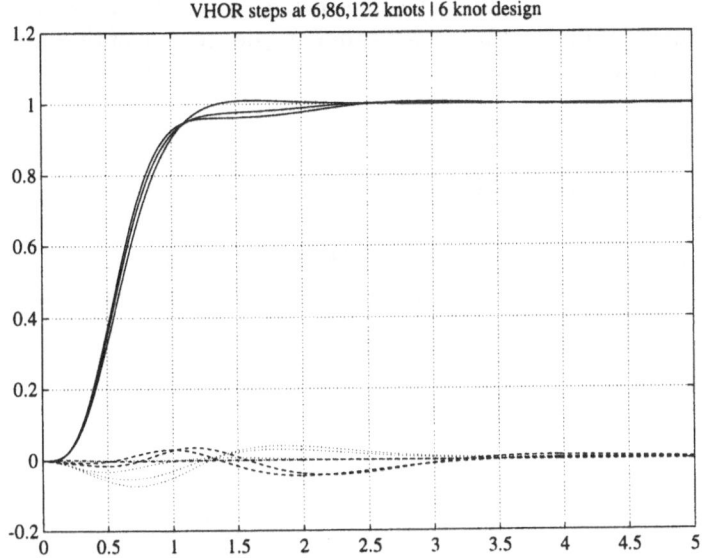

Fig. 4.17. 6 knot loop-shaping design — robust performance analysis.

$$0.8 \leq k_2 \leq 1.2$$

$$0.0 \leq \tau \leq 1.0$$

Units are minutes and radians/minute. The condition number of $G(s)$ is 142 at all frequencies. This arises because the two outputs to be controlled are highly coupled. The actuators have similar capabilities.

The performance requirement is that a step demand on either output should be above 90% after 30 minutes, and coupling to the other output should be less than 50%. Steady-state error should be zero.

When doing the loop-shaping design, the plant with $k_1 = 1.0$, $k_2 = 1.0$, and $\tau = 1.0$ is used. The reason for taking $\tau = 1.0$ is that an increase in the time-delay will reduce phase margins, whereas a decrease should improve phase margins for this particular plant. The time-delay is approximated with a first order Padé approximation for both design and simulation.

The design procedure

1. Output scaling. From the decoupling requirements it can be seen that this is already done as 50 % interaction is equally undesirable on each of the two outputs.
2. Input scaling. The two actuators are already appropriately scaled as both have similar capability (see [41] for a full physical description of the plant).

3. Re-order inputs and outputs. This does not make any difference for this example because both actuators have similar gains to each of the outputs.
4. Alter the roll-off rates. There is no need for this as all the singular value roll-off rates are already 20dB/decade. Integral action is added to give zero steady state error using

$$W_p = \begin{bmatrix} \frac{s+0.5}{s} & 0 \\ 0 & \frac{s+0.5}{s} \end{bmatrix}$$

The zeros at 0.5 are to ensure that at cross-over the roll-off rate is not too close to 40dB/decade. The shaped singular values are shown in 4.18.

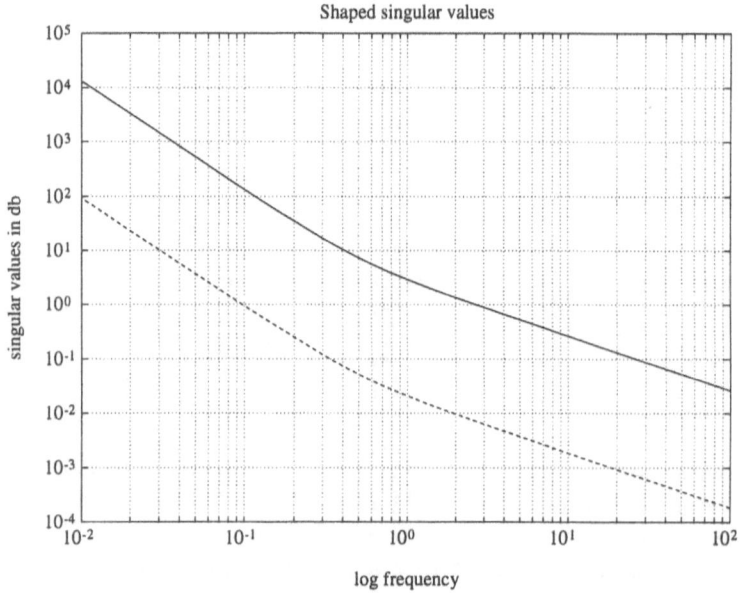

Fig. 4.18. Shaped singular values for the fractional distillation column.

5. Aligning. As the plant is so poorly conditioned, we do not align for the reasons discussed in §4.1.
6. Optimisation. The optimal γ_{min} is 3.88. It is interesting to note that γ_{min} is raised by the presence of the time-delay. If the design is re-run with no time-delay in the plant, the achieved γ is 2.84. Put $\gamma = 4.0$. Figure 4.19 shows the final open-loop singular values. It can be seen that K_∞ has altered the singular values, which is consistent with the relatively high γ value.
7. Figures 4.20 and 4.21 show superimposed step responses for different uncertainty combinations in the plant, including different time-delays.

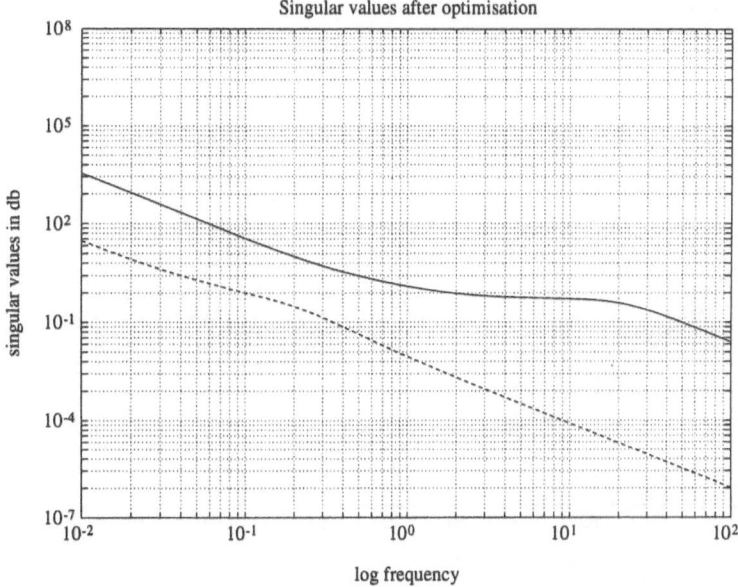

Fig. 4.19. Final singular values for the fractional distillation column.

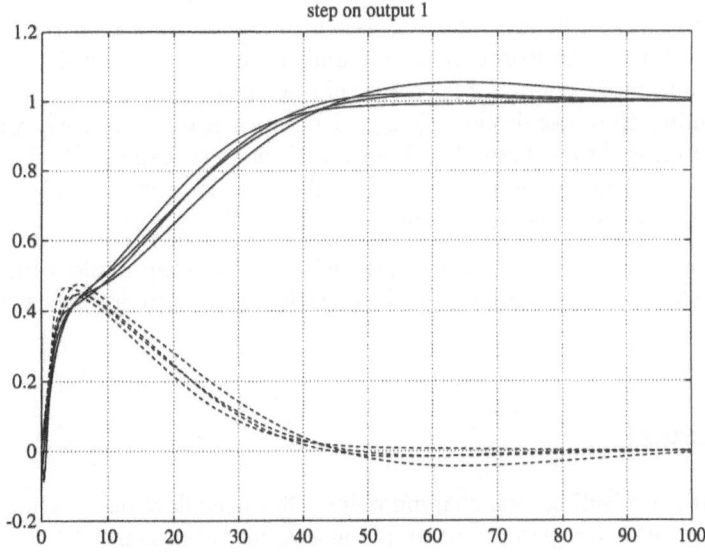

Fig. 4.20. Steps on output 1 for the fractional distillation column.

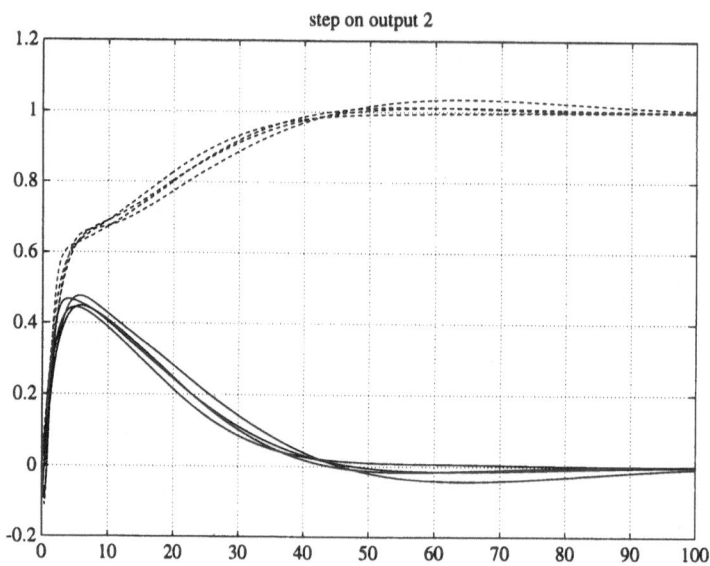

Fig. 4.21. Steps on output 2 for the fractional distillation column.

These compare well with the design in [47] in that overshoot is less, and interaction less. The loop-shaping design probably has a slightly higher bandwidth than the Horowitz design, and this could be reduced to make the results more comparable. Note that it is not possible to increase the bandwidth of the design any higher because of the time-delay which results in loss of robust stability. Figure 4.22 shows that the multiplicative input and output robustness are reasonable; -10dB corresponds to 30 % tolerable multiplicative uncertainty.

The design was re-run with no time-delay to be comparable with the structured singular value design in [41]. Again it was easy to produce similar results.

4.8 Summary

A procedure for selecting loop-shaping weights has been developed. This procedure has proved straightforward to use on a variety of design examples, including the ones presented here. One of the key strengths of the loop-shaping approach is that it is usually clear how to alter the weighting functions if the performance of the initial design is not quite right. Combined with the fast optimisation solution which does not require an iteration to find γ_{min}, a very

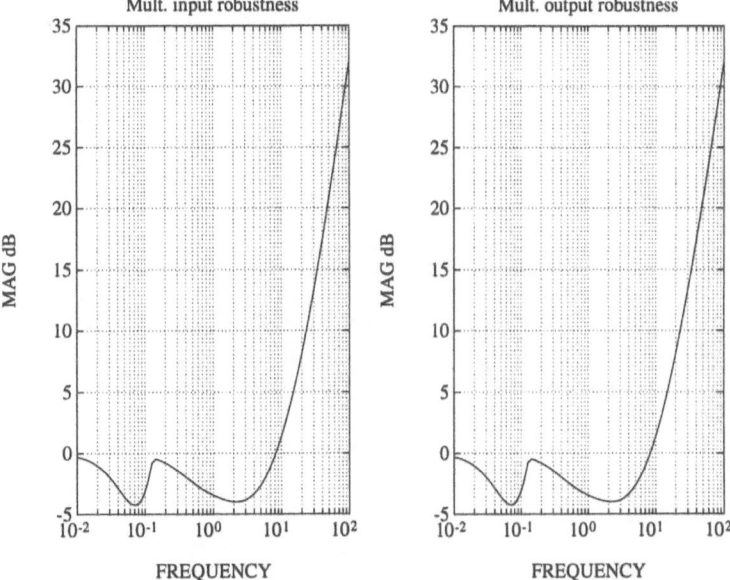

Fig. 4.22. Distillation column design multiplicative input and output robustness.

powerful design tool results. Although not part of the explicit design formulation, on all the designs presented here robust performance has been good. This is a very important characteristic for in that it reduces the number of design points needed to cover the operating envelope.

CHAPTER 5
CONTROLLER SWITCHING

5.1 Introduction

The idea of capturing the non-linear nature of the plant using key parameters has already been introduced. Having formed a set of linear controllers designed on linearisations of the plant for a set of values of these key parameters which cover the operating envelope, some way of moving between the controllers as the operating envelope is traversed is required. One possible approach is to assign subsets of the operating envelope to each linear controller, and then to switch between them as the boundaries between the regions are traversed. The GVAM design example presented here uses only one key parameter, namely airspeed. Each of the linear controller designs is therefore assigned to a particular airspeed range.

Switching from one controller to another is not generally a straightforward task. The problem arises of how to initialise the states of a controller when it is brought on-line; inconsistent initialisation of the controller states would lead to step changes on the plant actuator demands which is clearly highly undesirable. There is also the possibility that the initial state could affect the system such that the key parameter changes so as to take the system back to the previous controller, and thereby possibly initiate an instability. One approach to this state initialisation problem has been suggested by Hanus [6], and effects so-called "bumpless transfer" between controllers. The term "bumpless" is used because no discontinuity appears directly at the plant input with this approach. The approach is described in §5.2, and its merits and limitations discussed.

Flight control systems present a particularly tough design specification for switching; the pilot desires a highly predictable response from the controlled aircraft, and any perceivable transient effect on actuators or aircraft state would detract from any design. The military specification MIL8785C states that:

"The transient motions and trim changes resulting from the intentional engagement of disengagement of any portion of the primary flight control system by the pilot shall be such that dangerous flying qualities never result. .... With the controls free, the transients shall not exceed ±0.1g normal or lateral acceleration at the pilot's station, and ±3 degrees per second roll."

These requirements are intended for mode changes directly initiated by the pilot; switching between controllers as described above is not directly initiated by the pilot, and so much more stringent requirements might be necessary.

The Hanus approach is illustrated by a design on the GVAM in §5.3. The stability of the technique is then discussed in §5.4, and a simple but conservative test for stability developed. A summary of the chapter's main results and conclusions are then given in §5.5.

5.2 Bumpless transfer using the Hanus approach

5.2.1 The Hanus approach

The Hanus technique in [6] achieves smooth transition between controllers by ensuring that the states of the next controller to be switched in are always consistent with current plant states. The set-up is shown in Figure 5.1

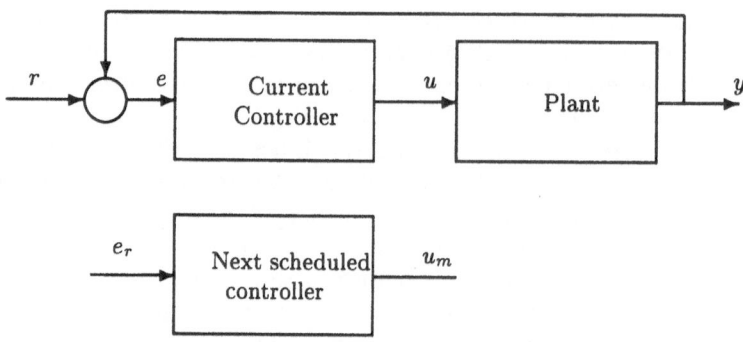

Fig. 5.1. Hanus anti-windup and bumpless transfer technique.

The e_r for each off-line controller is the error signal which would have given present plant inputs had that controller been the current on-line one. If the states of a particular linear off-line controller are denoted by x_h, then the output equation is :

$$u = C_K x_h + D_K e_r.$$

Given u and x_h it is possible to calculate e_r provided D is invertible:

$$e_r = D_K^{-1}(u_m - C_K x_h)$$

Substituting into the state equation gives

$$\dot{x}_h = (A_K - B_K D_K^{-1} C_K x_h + B_K D_K^{-1} u_m$$

Using the state-space notation of [23], the controller can be written in the self-conditioned form

$$u = \left[\begin{array}{c|cc} A_K - B_K D_K^{-1} C_K & O & B_K D_K^{-1} \\ \hline C_K & D_K & O \end{array}\right] \left[\begin{array}{c} e \\ u_m \end{array}\right]$$

This self-conditioned form is always driven by current plant inputs, u_m, and the loop error signal, e, whether the controller is on-line or not. The controller output, u, is only used when the controller is on-line.

5.2.2 Implementation of loop-shaping controllers

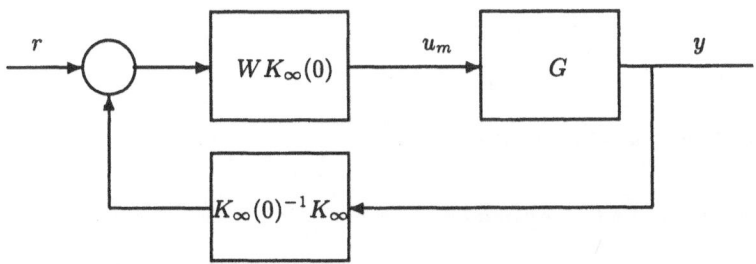

Fig. 5.2. Implementation of loop-shaping controllers.

Controllers from the loop-shaping procedure are implemented as shown in figure 5.2 where W is the precompensator chosen by the designer, K_∞ is the controller from the optimisation procedure, and $K_\infty(0)$ is K_∞ evaluated at $\omega = 0$. As already discussed, the motive for this particular structure is to avoid directly exciting the dynamics of K_∞ with the references. Note that no postcompensator W_2 has been included as this can just be cascaded with K_∞ before implementation. Hence when the controller is on-line we have

$$u = W K_\infty(0) r - W K_\infty y$$

and when being conditioned off-line

$$y_r = K_\infty^{-1} K_\infty(0) r - K_\infty^{-1} W^{-1} u_m$$

If we write $W K_\infty(0)$ and $K_\infty(0)^{-1} K_\infty$ in state-space form as:

$$W K_\infty(0) \equiv \left[\begin{array}{c|c} A_W & B_W \\ \hline C_W & D_W \end{array}\right]$$

and

$$K_\infty(0)^{-1} K_\infty \equiv \left[\begin{array}{c|c} A_\infty & B_\infty \\ \hline C_\infty & D_\infty \end{array}\right]$$

then it is easy to show that both the on-line and off-line cases can be implemented by using the self-conditioned form:

$$u = \left[\begin{array}{c|c} A_T & B_T \\ \hline C_T & D_T \end{array}\right] \left[\begin{array}{c} r \\ y \\ u_m \end{array}\right]$$

where

$$A_T = \left[\begin{array}{cc} A_W - B_W D_W^{-1} C_W & 0 \\ -B_\infty D_\infty^{-1} C_W & A_\infty - B_\infty D_\infty^{-1} C_\infty \end{array}\right]$$

$$B_T = \left[\begin{array}{ccc} 0 & 0 & -B_W D_W^{-1} \\ B_\infty D_\infty^{-1} & 0 & -B_\infty D_\infty^{-1} D_W^{-1} \end{array}\right]$$

$$C_T = \left[\begin{array}{cc} -C_W & -D_W C_\infty \end{array}\right]$$

$$D_T = \left[\begin{array}{ccc} D_W & -D_W D_\infty & 0 \end{array}\right]$$

and $u_m = u$ if the controller is on-line.

5.2.3 Limitations of the approach

It is clear that for the Hanus approach to be applied, the linear controllers must be:

1. Invertible.
2. Minimum phase.

When implementing a controller, a constant term (i.e. a non-zero D-matrix) can be undesirable because it will allow high frequency noise on the measurements to feedback on to the control action. A possible solution to this is to have a postcompensator common to all of the linear controllers which attenuates high frequencies as illustrated in Figure 5.3:

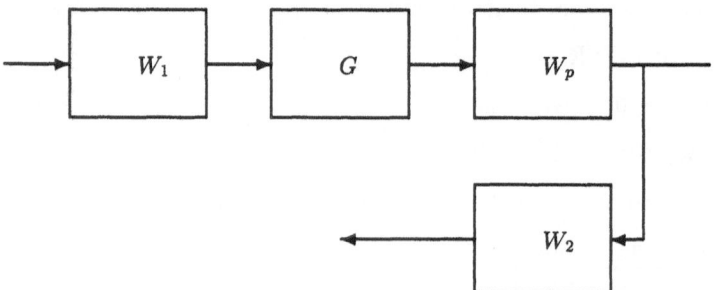

Fig. 5.3. Use of W_p to attenuate high frequency noise.

The plant is then taken to be W_pG for design, and for implementation of the Hanus strategy.

The minimum phase condition is unlikely to cause problems for most design problems. However, the speed of the zeros is also important; if the inverted off-line controller has much slower dynamics than the dynamics of the closed-loop effected by the current on-line controller, then the controller states may not be consistent at the time of the controller switch over. Also, the faster the zeros of the controller, the closer to the switching boundary the system can be before initialisation of the next controller has to commence. This has computational advantages, particularly if switching is done on several key parameters. Hence the dynamics of W_1 and W_2 should be chosen with this in mind.

5.2.4 Selecting the 'D'-matrix

For a general $\mathcal{H}_\infty$ problem, all controllers K_∞ for which $\|\mathcal{F}_L(P,K)\|_\infty < \gamma$ are parameterised [2] by $\mathcal{F}_L(M_\infty, Q)$ where M_∞ has the form

$$M_\infty = \left[\begin{array}{c|cc} \hat{A} & \hat{B}_1 & \hat{B}_2 \\ \hline \tilde{C}_1 & 0 & I \\ \hat{C}_2 & I & 0 \end{array} \right]$$

where $Q \in R\mathcal{H}_\infty$, and $\|Q\|_\infty < \gamma$.

The minimum phase condition required for scheduling is equivalent to K^{-1} being stable. It is relatively straightforward to show that [48] all K^{-1} are parametrised by $\mathcal{F}_L(E_\infty, Q^{-1})$ where

$$E_\infty = \left[\begin{array}{c|cc} \hat{A} - \hat{B}_1\hat{C}_2 - \hat{B}_2\hat{C}_1 & \hat{B}_2 & \hat{B}_1 \\ \hline -\hat{C}_2 & 0 & I \\ -\hat{C}_1 & I & 0 \end{array} \right]$$

Hence the problem is equivalent to finding Q^{-1} which stabilises E_∞ with the restrictions that $Q \in R\mathcal{H}_\infty$ and $\|Q\|_\infty < \gamma$. Hence finding a suitable Q is equivalent to finding a stabilising Q^{-1} for the system

$$V_\infty = \left[\begin{array}{c|c} \hat{A} - \hat{B}_1\hat{C}_2 - \hat{B}_2\hat{C}_1 & \hat{B}_1 \\ \hline -\hat{C}_1 & 0 \end{array} \right]$$

where $\|Q\|_\infty < \gamma$, and Q^{-1} has no right half plane zeros. This can be treated as a characteristic locus design problem.

In practice it has been found to be very straightforward to select a Q which gives a minimum phase solution; for all the GVAM designs, $Q = I_3$ sufficed, and hence $D = I_3$. It has also been noted that Q has very little effect on the slowest zeros of K_∞, and it was therefore not found possible to have a significant effect on the conditioning time of the off-line controllers through the selection of Q.

5.2.5 Relationship to the high gain approach

An alternative approach to initialising the states of the controller is to use what is referred to here as the "high gain antiwindup approach". The method is illustrated in Figure 5.4.

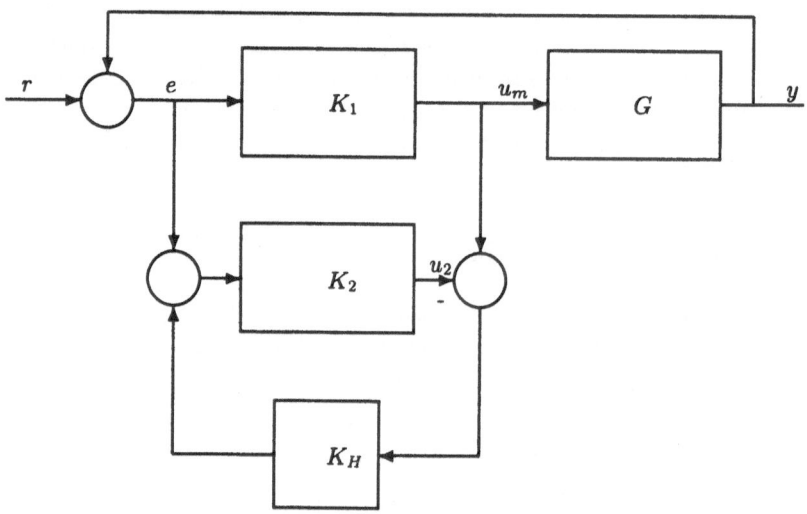

Fig. 5.4. High gain approach to controller conditioning.

Controller K_1 is on-line, and K_2 is being conditioned off-line. K_H is a high gain (possibly constant) transfer function which alters the input to K_2 thereby forcing u_2 to track u_m. Writing the state equation for K_2 we get:

$$\dot{x} = (A_2 - B_2 K_H C_2)x + B_2 K_H u_m + B_2 e$$

If we put $K_H = D_2^{-1}$, then we get

$$\dot{x} = (A_2 - B_2 D_2^{-1} C_2)x + B_2 D_2^{-1} u_m + B_2 e$$

Note that this is not identical to the Hanus approach since we have an extra term, namely $B_2 e$. However, if D_2 is made small, then it does tend to the Hanus configuration.

 This high gain configuration has the advantage that it allows more freedom in the selection of the zeros of the controller i.e. poles of $A_2 + B_2 K_H C_2$. It also does not require the linear controllers to have constant terms. These benefits are at the expense of a larger "bump" when transferring between controllers.

5.3 GVAM switched controller example

5.3.1 The linear designs

The switching design presented here is the one which was simulated in the
R.A.E. simulator in April 1991 [4] Four linear designs based on linearisations
of the GVAM at 6, 86, 140 and 170 knots airspeed were used to cover the
flight envelope. In deciding how close the design points need to be, both robust
stability margins and step responses were used. It was found that switching
on just forward airspeed was all that was required; the linear controllers
exhibit sufficient robust performance for switching on angle of attack not to
be required for normal flying i.e. incidence angles up to 15 degrees.

The speed range 0—200 knots is where the dynamics vary most significantly. For speeds above 200 knots the nozzles are fully aft, and the dynamic
variation with speed is less marked. The 6 knot design is that detailed in chapter 4. The other designs were carried out using exactly the same methodology.

Only the switching between the first three controller designs is discussed
here, as switching to and from the 170 knot controller involves a mode change
as well in that the nozzles are always demanded aft by the fourth controller.
A 5 knot hysteresis was put in all the switching speeds; this is necessary in
the presence of sensor noise and air turbulence to avoid unnecessary rapid
switching between controllers.

5.3.2 Linear analysis

An obvious way to evaluate the performance of the switched controllers across
the flight envelope is to plot out the achieved $\mathcal{H}_\infty$ cost function used for the
linear designs. Figure 5.5 shows the achieved normalised left coprime factor
robustness optimisation cost, γ, as a function of airspeed for the 6 and 86
knot designs (i.e. 10 and 145 feet/s designs).

As might be expected γ rises as the airspeed moves away from the design
points. Deciding how large a γ can be tolerated can be used to determine
how close the design points need to be.

Figures 5.6, 5.7 and 5.8 show superimposed time responses for step demands on VHOR, VKD and THETD for various airspeeds. All of these responses meet the required handling qualities given in chapter 3 when suitable
command types and scaling are used.

5.3.3 Implementation in TSIM

The linear controllers were implemented on the non-linear model using the
self-conditioned Hanus structure implemented in TSIM. Initialisation of the
controller states at the start of the simulation is easy; the actuators are just
held at the hover stop values for two seconds, and then the controller is
allowed to take over. This is a great advantage of Hanus approach in that the

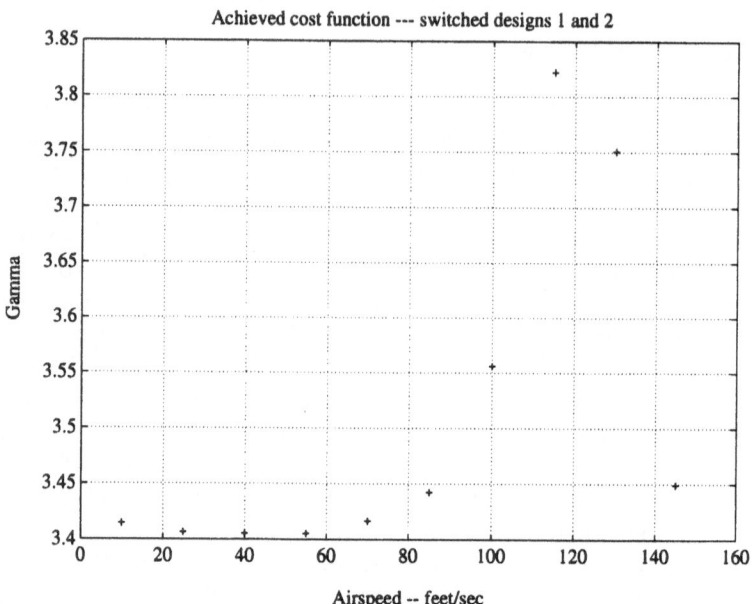

Fig. 5.5. Achieved γ between K_1 and K_2 design points, switch at 120 knots

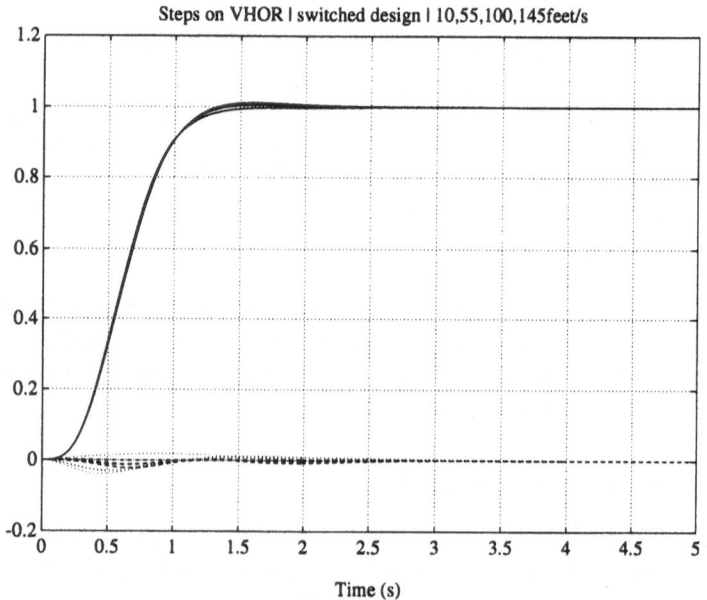

Fig. 5.6. Forward speed step responses

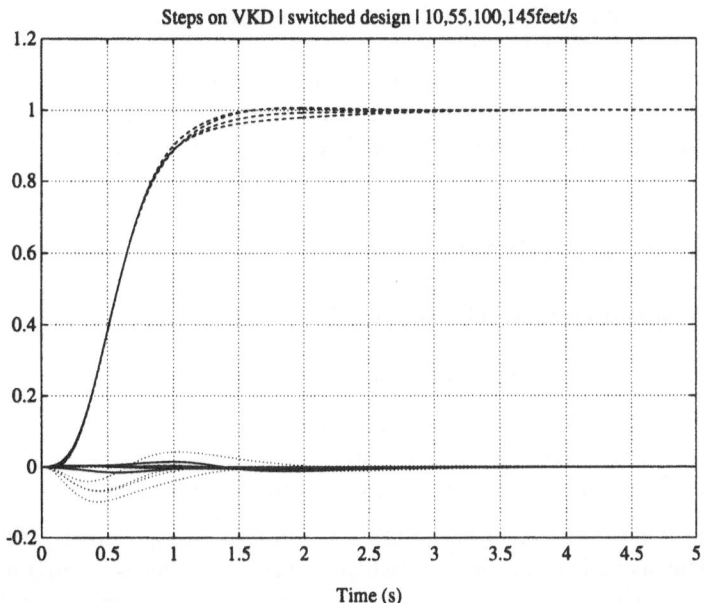

Fig. 5.7. Vertical speed step responses

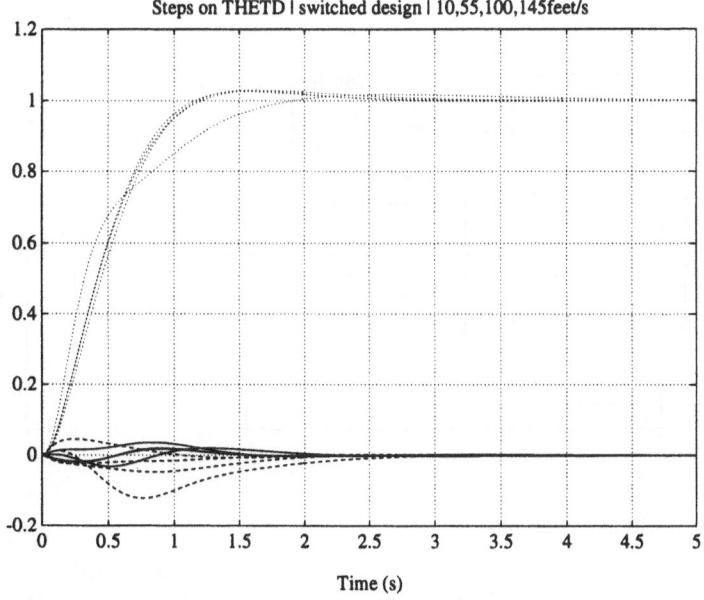

Fig. 5.8. Pitch attitude step responses

automatic control system can be brought on or off-line at any stage in flight without resorting to lookup tables for state initialisation. The controllers were implemented in discrete form using a sample period of 20 milliseconds.

The results of the piloted simulation can be found in [4]. The main result was that despite the "bumpless" transfer, the pilot could often detect the switching points due to the discrete change in performance.

5.4 Stability analysis

5.4.1 A stability analysis technique for switching

Obvious necessary conditions for stability are :

1. the on-line controller gives an asymptotically stable closed-loop for asymptotic reference demands, and
2. the off-line controller states must tend to a steady state for a given u.

The first condition can be ensured by making the design points sufficiently close. The second condition is satisfied if each controller is minimum phase. Notice that the technique is essentially inverting the off-line controller and running it backwards. Hence if the zeros of K can be made fast as well as minimum phase it will become consistent with the plant states quickly. If the inverted controllers are sufficiently fast, then in a scheduling regime only the next scheduled controller need be updated in addition to the current controller.

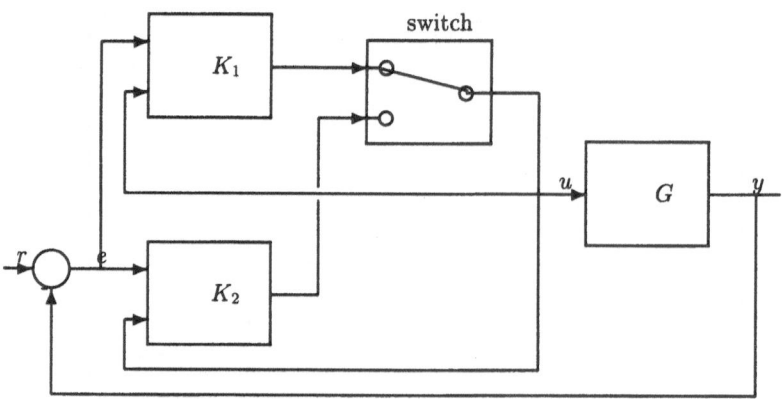

Fig. 5.9. Switched controller analysis.

Consider the set-up in Figure 5.9. It shows two controllers, both of which are updated using the self-conditioned Hanus structure. All that is required to effect the switch when using this implementation is to select the output of either K_1 or K_2 as the input to the plant. The switch can now be modelled as a non-linear time-varying gain, λI, as in Figure 5.10(a). $\lambda = 0$ corresponds to K_2 on-line and $\lambda = 1$ corresponds to K_1 being on line. The block diagram of Figure 5.9 can then be redrawn in the form of Figure 5.10(b) where the $\Delta = \lambda I$.

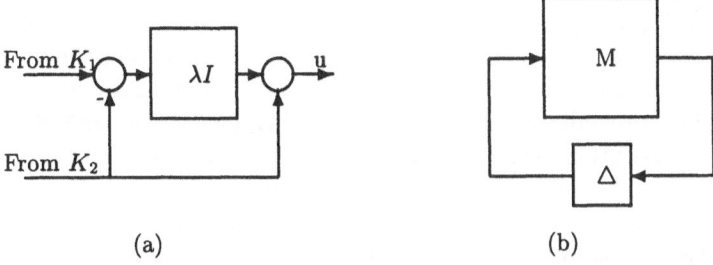

(a)   (b)

Fig. 5.10. Switching stability test using the structured singular value.

Using the small gain theorem for M linear time-invariant and Δ non-linear, memoryless and time-varying [12] the system is stable if

1. M is stable , and
2. $\|M\|_\infty < 1$

The above test however may be very conservative since the small gain theorem considers Δ as norm bounded by 1, whereas here we have $\Delta = \lambda I$ where λ is real valued between 0 and 1. The diagonal structure of the Δ can be accounted for by replacing (2) with the test

$$\inf_D \left\| DMD^{-1} \right\|_\infty < 1$$

where D is a constant invertible matrix.

The time-varying nature of the plant has not been taken into account. It is likely that the aircraft will cross the switching line whilst in an acceleration regime. To evaluate stability further, the above tests could be applied to linearisations of the plant at acceleration operating points. Alternatively the effect of acceleration could be modelled as real perturbations to the A,B,C and D matrices, and the test reposed as a structured singular value type problem with two repeated blocks and constant D matrices. Note that one might expect less of a problem during an acceleration regime as the aircraft will tend to go straight past the switching speed.

This gives a way of evaluating stability at the switch. Due to the conservative nature of the test it may not always be of use. Also it only evaluates stability, and does not guarantee that as the switching line is traversed that no oscillation between controllers will occur. Such oscillation may be undesirable in that it could excite high frequency dynamics of the airframe, and also give undesirable responses to pilot demands. To prevent this some form of hysteresis needs to be built into the switching speed. The size of the hysteresis required is dictated by :

1. noise levels on the speed signal, and
2. the dynamic effect of the switch on forward speed.

To establish the size of hysteresis required due to the switching dynamics, time simulations of various manoeuvres are performed. With the designs presented here it was found that no hysteresis was required in the absence of noise. In the absence of available data on the noise levels on the speed measurement, a nominal 5 knot hysteresis was used for the design study.

This stability test falls into the framework of [49],[50] for analysis of limitations and substitutions where the isolation of non-linearities using conic sectors is proposed. Conic sector bounding is defined by:

Definition : (Campo *et al*) Given $N : L_{2e} \rightarrow L_{2e}$ and the LTI operators C and R, N is said to be inside Cone(C,R) if

$$\|N(x) - Cx\|_T \leq \|Rx\|_T$$

for all $T \geq 0$ and for all $x \in L_{2e}$, where L_{2e} is the extended space of vector valued functions, $x(t)$, with the property

$$\|x(t)\|_T \equiv [\int_0^T x^*(t)x(t)dt]^{1/2} < \infty$$

for all $T \geq 0$.

For the nonlinearity used above in the switching analysis we have R=I and C=0.

5.4.2 Application to the GVAM Example

Figures 5.11 and 5.12 show the results of the stability test for switching between controllers 1 and 2, and controllers 2 and 3 respectively. Recall that this singular value should be less than one if stability for the two switched controllers on the linearised plant model is to be predicted. The structured singular value (μ) which treats λ as being non-timevarying and complex is also plotted for comparison. To find an approximate solution to the constant D restriction for the time-varying case, the approach was simply to take the optimal D-scale at the frequency for which μ was a maximum. This approach may therefore give a conservative result. Figure 5.11 suggests that there will be no problem switching between controllers 1 and 2. However, Figure 5.12 shows that the test is too conservative for switching between controllers 2 and 3. This suggests that the switch between controllers 2 and 3 may be more noticeable. This was indeed found to be the case.

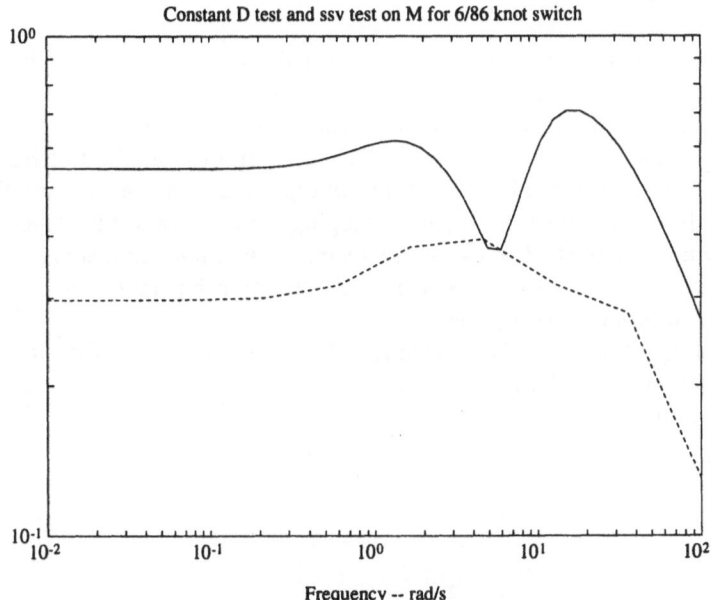

Fig. 5.11. Switching stability test — designs 1 and 2

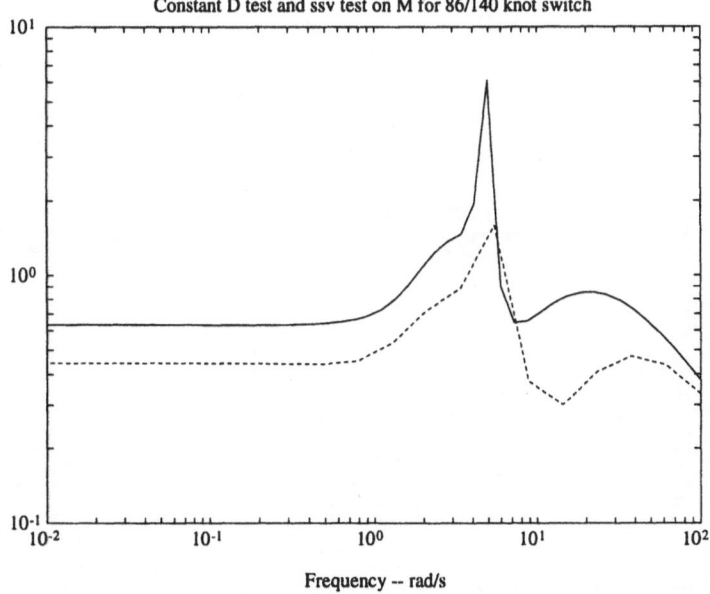

Fig. 5.12. Switching stability test — designs 2 and 3

5.5 Summary

The idea of switching between linear controller designs has been introduced. The approach allows non-linearity captured by key variables to be accounted for in the control law. Having motivated the idea of switching, the practicalities of bumpless transfer have been discussed, and the method of Hanus introduced. It has be shown how the Hanus bumpless transfer can be modified for use with linear designs from the loop-shaping approach, and the technique successfully applied to the GVAM design example. Possible limitations of the Hanus method have been discussed, and it's relationship to the high gain anti-windup approach investigated.

A switching stability analysis technique has been proposed. However the test is conservative, and time domain simulation is required to check the performance qualities anyway.

CHAPTER 6
CONTROLLER SCHEDULING

6.1 Introduction

Having motivated the idea of capturing non-linearity with key parameters, the idea of switching between controllers has been introduced in chapter 5. In this chapter an alternative approach is used which involves making the control law a continuous function of the key (scheduling) parameters. The approach involves obtaining a set of linearisations for a set of values of the scheduling parameter(s), designing a set of linear optimal controllers, and then linearly interpolating the linear controller gains as the operating envelope is traversed.

This approach is consistent with existing control laws which typically schedule proportional and integral gains with operating point. As yet there has been very little application of scheduled optimal controllers. This is because the linear controllers must all have the same structure in order that gains may be interpolated. Conventional P+I controllers obviously have a clearly defined structure, and gain scheduling multivariable P+I controllers is also very straightforward - see for example [51]. LQG controllers have been scheduled using the inherent plant observer structure of such controllers - see for example [52]. $\mathcal{H}_\infty$ controllers in general do not have such an explicit structure, and scheduling presents a problem. However, the normalised co-prime factor robust stabilisation approach does produce a controller which can be written as a plant observer, and this property is used to implement the designs in this chapter. Some progress has also been made by Kellet [53] for scheduling controllers obtained from $\mathcal{H}_\infty$ closed-loop optimisation by using balanced realisations.

Gain scheduling is a widely accepted technique, and is routinely used in many control applications. There is, however, very little in the way of a theoretical foundation for scheduling. In particular there is a lack of tests which can be applied to guarantee stability of scheduled systems, let alone to evaluate performance. Designers have to rely on extensive time-simulations to evaluate stability and performance; apart from being heavy on human resources, this still does not guarantee anything. Shamma and Athans have recently made some progress in this area. Their initial work in concentrated on linear parameter-varying plants, and they later extended the work to non-linear plants [54], [55]. In their analysis the scheduled non-linear system is written as a Volterra integro-differential equation, and Lyapunov functionals are used to get conditions for exponential stability. A condition is given for robustness and performance properties of the frozen operating points on the schedule to carry over to the full gain scheduled system. In practice the approach is hard to apply, and apt to be conservative. However, a couple of very useful properties of gain scheduled systems are formalised by their results. Firstly, the stability margin is proportional to the level of frozen operating point stability. This provides further motivation for using robust linear controller designs. Secondly, the stability margin is inversely proportional to the rate of change of the scheduling variable. This gives a theoretical justification to the rule of thumb that the scheduling variable should vary slowly relative to the closed-loop dynamics.

An interesting comment made in [55] is that the closed-loop "dynamics matrix" arrived at from writing the schedule as a linear volterra integro-differential equation [55, Equation 21] differs from that obtained from applying the linear controller to the plant linearised at a frozen operating point on the schedule. This leads to the proposition that instead of designing the controllers for linearisations corresponding to constant values of the scheduling variable, perhaps they should be done on linearisations for constant values of the rate of change of the scheduling variable.

Here an analysis approach using the structured singular value is developed. Again the approach is highly conservative, but the framework of the analysis has potential in the future for development of less conservative tests.

Section 6.2 describes the scheduling of $\mathcal{H}_\infty$/loop-shaping controllers, and following this a couple of design examples are presented in §6.3. In §6.4 the structured singular value scheduling analysis approach is developed and applied to one of the two design examples. Finally the results and contribution of the chapter are summarised in 6.5.

6.2 Scheduling using an observer structure

6.2.1 Writing loop-shaping controllers as observers

In [16] it was shown that the controller resulting from loop-shaping approach can be written as an exact plant observer plus state feedback:

$$\dot{\hat{x}} = A\hat{x} + H(C\hat{x} - y) + Bu$$

$$u = F\hat{x}$$

where [A,B,C] is a state-space realisation of the weighted plant, $H = -ZC^*$, and $F = B'(\gamma^{-2}I + \gamma^{-2}XZ - I)^{-1}X$ where X and Z are the associated control and filtering algebraic Riccati equation solutions.

In general $\mathcal{H}_\infty$ controllers cannot be written as exact plant state observers as there will be a worst disturbance term entering the observer state equation as shown in [2]. However, for the loop-shaping controllers it is possible, and the clear structure lends itself to gain scheduling in that the F and H gains can be scheduled as a function of aircraft forward speed. Figure 6.1 shows a block diagram implementation of this.

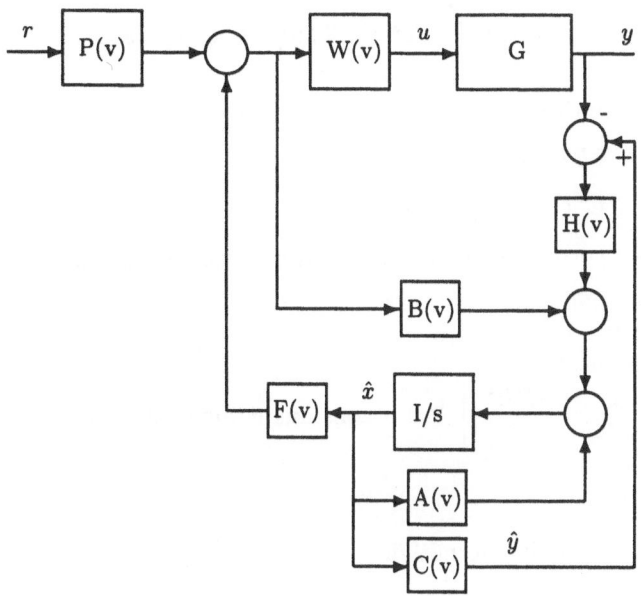

Fig. 6.1. An observer implementation of an $\mathcal{H}_\infty$/loop-shaping controller.

In order to be able to interpolate between controller gains we must have:

1. the shaped plant matrices A,B,C varying smoothly with operating point, and
2. the controller F and H gains varying smoothly with operating point.

Knowledge of the physical system to be controlled should be used to check that the first condition will not be violated. It can be seen that the second condition will be satisfied provided that the Riccati solutions X and Z and the γ vary smoothly as A,B and C vary smoothly. The following theorem from [56] can be used to show that this is indeed the case:

Theorem 6.2.1. *Let $A(\theta), D(\theta), C(\theta)$ be analytic $n \times n$ matrix functions on θ defined on a real interval (α, β), with $D(\theta)$ positive semidefinite Hermitian, $C(\theta)$ Hermitian, and $(A(\theta), D(\theta))$ stabilisable for every $\theta \in (\alpha, \beta)$. Assume that for all $\theta \in (\alpha, \beta)$ the Riccati equation*

$$X(\theta)D(\theta)X(\theta) - X(\theta)A(\theta) - A(\theta)^* X(\theta) - C(\theta) = 0$$

has a Hermitian solution. Further assume that the number of pure imaginary eigenvalues (counting multiplicities) of

$$M(\theta) = \begin{bmatrix} A(\theta) & -D(\theta) \\ -C(\theta) & -A(\theta)^* \end{bmatrix}$$

is constant. Then the maximal solution $X_m(\theta)$ is an analytic function of $\theta \in (\alpha, \beta)$.

Recall the CARE and FARE for the system $[A, B, C, 0]$:

$$A^*X + XA - XBB^*X + C^*C = 0$$

$$AZ + ZA^* - ZC^*CZ + BB^* = 0$$

For the CARE the corresponding assumptions in theorem 6.2.1 are:

- BB^* is positive semidefinite Hermitian.
- C^*C is Hermitian.
- (A, BB^*) is stabilisable.
- $M_C(t) = \begin{bmatrix} A & -BB^* \\ -C^*C & -A^* \end{bmatrix}$ has a constant number of pure imaginary eigenvalues.

and for the FARE,

- C^*C positive semidefinite Hermitian.
- BB^* is Hermitian.
- (A, C^*C) is stabilisable.
- $M_F(t) = \begin{bmatrix} A & -C^*C \\ -BB^* & -A^* \end{bmatrix}$ has a constant number of pure imaginary eigenvalues.

The Hermitian and positive semidefinite conditions are trivially satisfied. Stabilisability and detectability are standard assumptions for solution of the normalised coprime factor robust stabilisation problem. It follows from [2, Lemma 3] that M_C and M_F have no imaginary eigenvalues under the assumptions of (A, B) stabilisable and (C, A) detectable, and hence the last two conditions are satisfied. Hence if the linear design points are placed sufficiently close then interpolation of the parameters of A,B,C,F and H is possible. Note that we must ensure that we choose the plant weighting matrices to vary smoothly with operating point.

The approach taken when scheduling on the GVAM has been to use linear interpolation of gains between successive designs. So, for example, the F-matrix between adjacent design points i and j would be calculated as:

$$F(v) = (1 - v)F_i + vF_j$$

with $v = 0$ corresponding to operating point i, $v = 1$ corresponding to operating point j, and v varying linearly between them. An alternative would be to fit polynomials through all the F's for the whole schedule, but this increases the computation required for each evaluation of the controller.

6.2.2 Scheduling on several parameters

A double schedule (for example a schedule on airspeed (v) and incidence (α)) could be implemented thus:

$$F(v, \alpha) = (1 - v)\left[(1 - \alpha)F(0, 0) + \alpha F(0, 1)\right] + v\left[(1 - \alpha)F(1, 0) + \alpha F(1, 1)\right]$$

where $F(0, 0)$, $F(0, 1)$, $F(1, 0)$ and $F(1, 1)$ are the F-matrices at the design points. The great power of this scheduling approach is that as many of these scheduling corrections as required can be added, and this knowledge about the plant-nonlinearity is incorporated into the control law. For example, a third scheduling parameter might be aircraft weight which gradually decreases as the fuel is used up. Note that the dynamic order of the compensator is not increased when adding more scheduling parameters.

6.2.3 Limitations of the approach

Notice that the observer is for the weighted plant. The weighting matrices for the linear controller have to be selected with this in mind; they must vary smoothly with operating point, and as such must have a fixed structure so that they can be linearly interpolated. This places some limitations at the linear design stage.

The complexity in terms of the number of parameters which have to be stored and updated is relatively high; essentially a complete parametric representation of the plant is stored, plus all of the values for the F and H matrices across the flight envelope. In a sense the weighting matrices are stored twice, once as the weighting matrix itself, and once in the A,B, and C matrices of the observer. Another consideration is that not all the parameters should need scheduling - for example some actuator states might not depend on the scheduling parameters. An procedure for deciding which parameters do and do not need scheduling is proposed in the next section.

The precompensator, P, must also be scheduled to account for the change in the low frequency gain of the controller as the controller is scheduled. Notice that the references do not enter the loop at the same point as for the implementation structure proposed in chapter 4 where K_∞ is placed completely in the feedback path. The observer structure here has also been found to give good time-response properties, in particular not exhibiting any overshoot. There is of course no reason why the references should not be fed in at the input to the precompensator W, thus effectively putting K_∞ in the feedback path as is done for the switched designs in the previous chapter.

6.3 GVAM design examples

6.3.1 Scheduling on forward speed

Two scheduled observer controller designs are investigated for the hover through 120 knot speed range. The first one uses linear interpolation between the parameters of six controller designs, whereas the second uses linear interpolation between only two controller designs. The designs make use of the same weighting procedure as given in chapter 4 for the hover design example. The dynamic parts of the plant precompensator are identical for all designs, and it is the align matrix which varies from one design to the next. Hence scheduling the weights simply requires linear interpolation of the nine elements of the non-dynamic align matrix.

The most obvious way to check the performance of the scheduled controller is to look at the achieved $\mathcal{H}_\infty$ cost objective across the flight envelope. Since the $\mathcal{H}_\infty$ cost function for the loop-shaping approach used represents the tolerance to additive perturbations to the normalised coprime factors of the plant, this can be used to decide how close the linear design points need to be to ensure good robust stability margins. Figure 6.2 shows the achieved $\mathcal{H}_\infty$/loop-shaping cost function (γ) for the two control laws, where '*' represents the observer interpolation using 6 design points, and 'o' represents the observer interpolation using 2 design points. As might be expected the

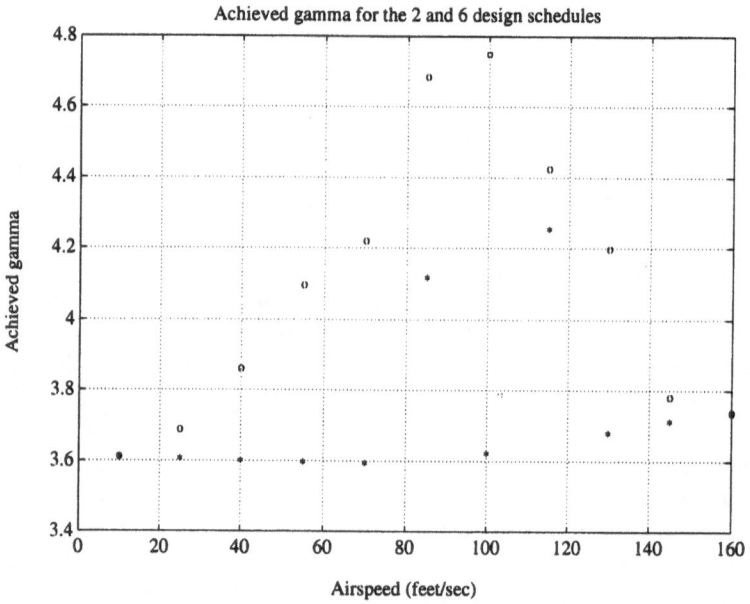

Fig. 6.2. Achieved $\mathcal{H}_\infty$/loop-shaping cost function, γ

controller using interpolation of six observers gives the best stability margins. Indeed, the achieved γ is very close to optimal over the whole speed range except around the 100feet/second point (approximately 60knots). In general this indicates that we should put in more design points around this speed.

However, this discontinuity is due to the opening and closing of the valve for the front pitching reaction jet as a function of the nozzle angle, and hence design using this particular linearisation is not appropriate.

Figures 6.3 and 6.4 show vertical speed demand step responses for the two control laws. The speed at which these are carried out is midway between the two designs of the 2-design controller schedule, and midway between two adjacent design points for the six-design controller. These plots indicate that, in terms of performance requirements, the schedule with just two designs is adequate.

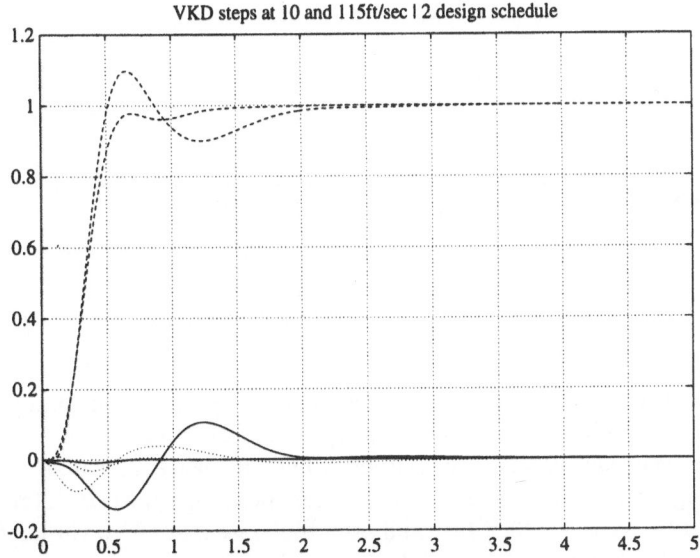

Fig. 6.3. Linear step responses for the two design point schedule

A schedule using three designs for the 0 to 140 knot speed range was implemented on the non-linear model. This control law was implemented on R.A.E. simulator, and the results of this can be found in [4]. This scheduling scheme worked better than the switching scheme presented in the previous chapter, and is used in Part II when designing a control law for flight testing.

6.3.2 Scheduling on speed and incidence

This design covers the operating range 130 to 160 feet/sec at angles of incidence between 8° and 40°. At these speeds the plant dynamics vary considerably with incidence as the aircraft is beginning to have appreciable aero-

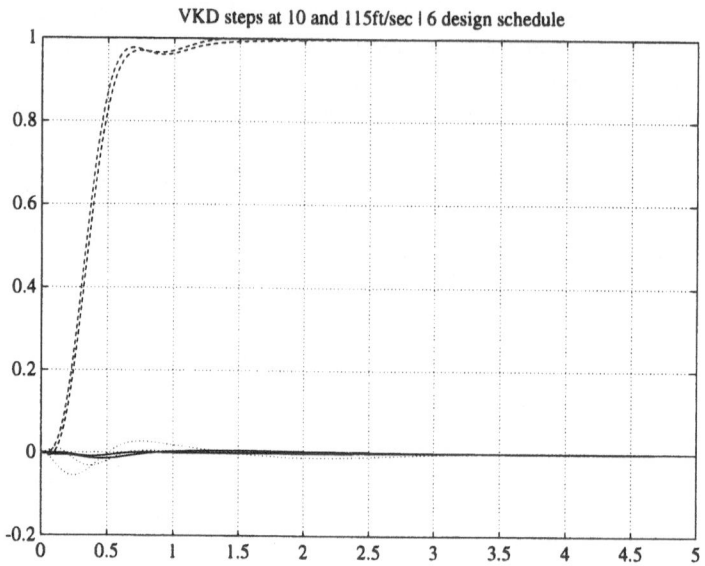

Fig. 6.4. Linear step responses for the six design point schedule

dynamic lift. Figures 6.5, 6.6 and 6.7 show forward speed step demands for a fixed linear controller designed at 160feet/sec and at 8° incidence.

It can be seen that performance is poor at high incidence. A double schedule with design points at the four corners of the operating space i.e. 130ft/sec and 8°, 160ft/sec and 8°, 130ft/sec and 40°, 160ft/sec and 40° was therefore carried out. The designs were scheduled using the double schedule approach outlined in §6.2.2. Table 6.1 shows the achieved γ at various points in the schedule, and indicates that robustness is good.

Table 6.1 — Achieved $\mathcal{H}_\infty$ norm					
Plant	$\alpha = 8^o$	$\alpha = 16^o$	$\alpha = 24^o$	$\alpha = 32^o$	$\alpha = 40^o$
vt=130ft/sec	3.86	4.72	4.92	4.47	4.18
vt=140ft/sec	4.35	4.51	4.77	4.41	4.31
vt=150ft/sec	4.22	4.35	4.66	4.41	4.74
vt=160ft/sec	3.99	5.20	4.78	4.45	4.44

Figures 6.8, 6.9 and 6.10 show forward speed step responses at various points in the schedule, and all these meet the required handling qualities.

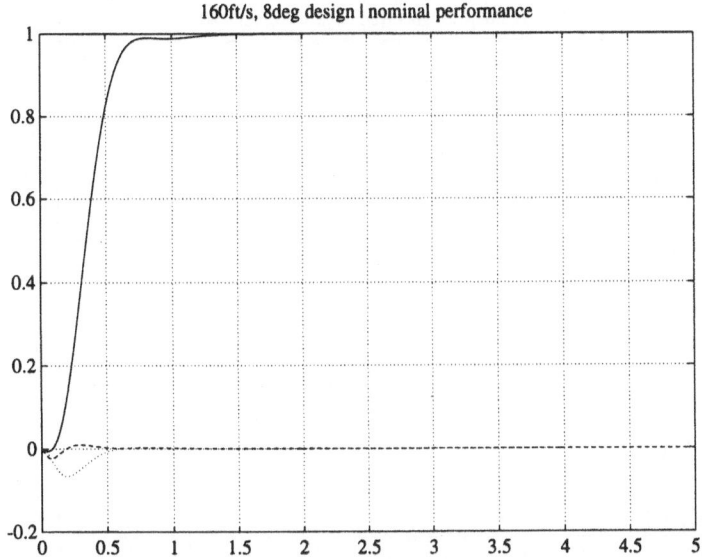

Fig. 6.5. Forward speed step response at design point (8° incidence)

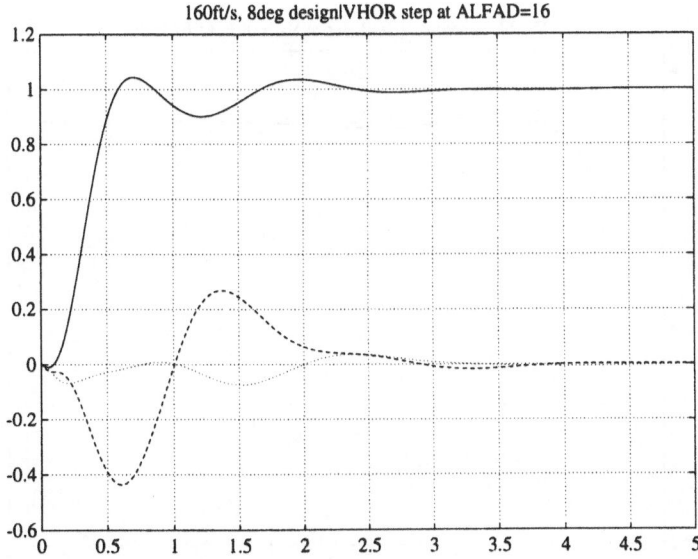

Fig. 6.6. Forward speed step response at 16° incidence

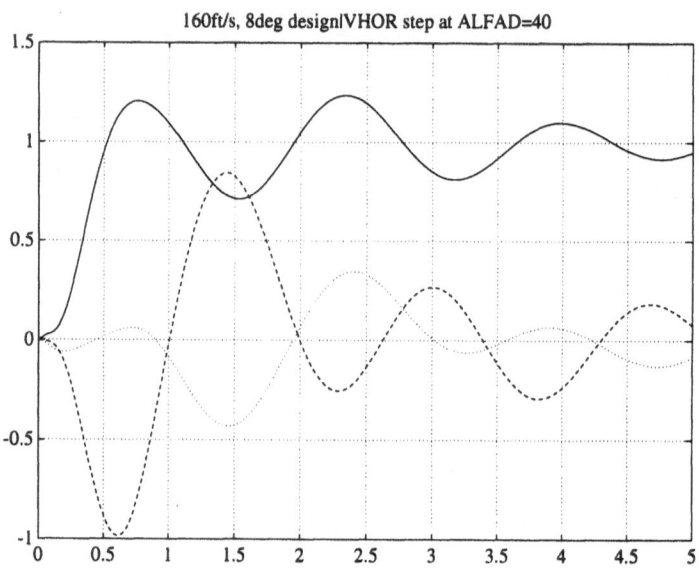

Fig. 6.7. Forward speed step response at 40° incidence

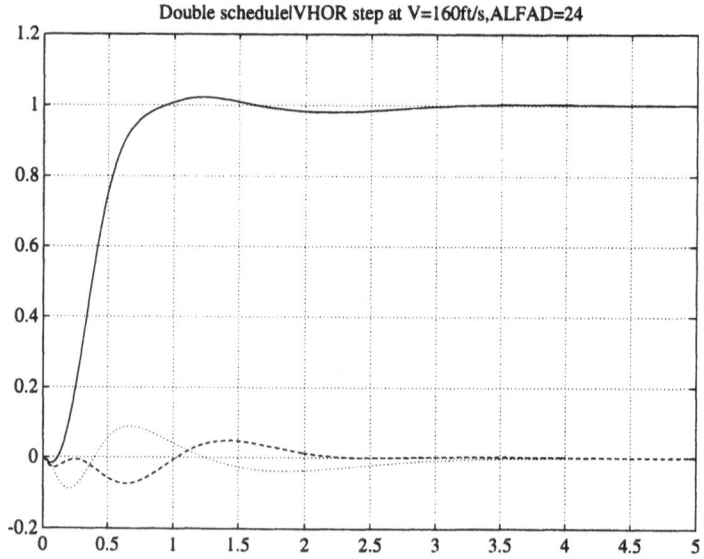

Fig. 6.8. Forward speed step for the scheduled controller

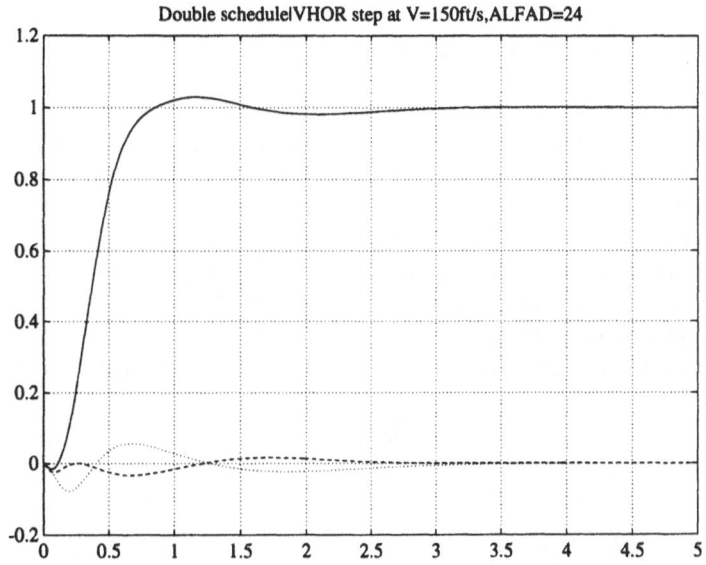

Fig. 6.9. Forward speed step for the scheduled controller

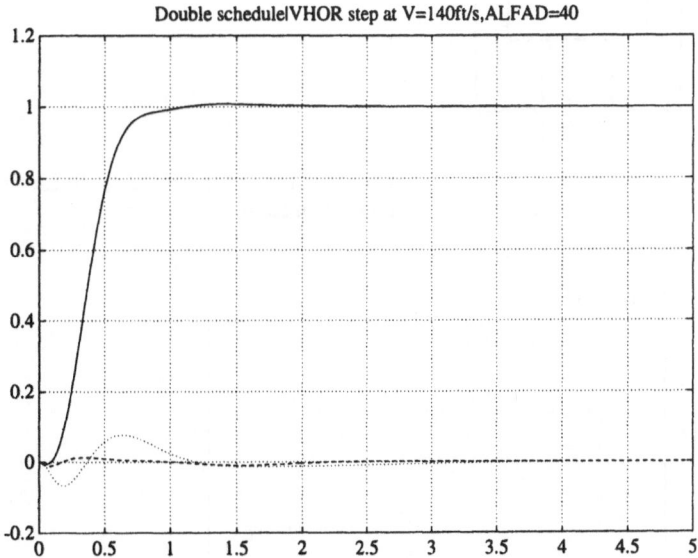

Fig. 6.10. Forward speed step for the scheduled controller

6.4 A linear analysis technique for scheduled controllers

6.4.1 The analysis procedure

When the controller is scheduled on several variables the parameter space that must be checked becomes very large. There is potential here to make use of the structured singular value, μ. One way to do this is to approximate the plant and and controller by Linear Fractional Transformations (LFT's). For the plant this can be achieved by fitting polynomials through the coefficients of the (A,B,C,D) state-space matrices as they change with the scheduling variable. For the linearisations used from the GVAM the C-matrix is constant, and the D-matrix is zero. Hence we only need to consider changes in the A and B-matrices. If pth order polynomials are fitted through the individual coefficients of $A_1 \dots A_n$ and $B_1 \dots B_n$ and the coefficients stored in matrices $\alpha_0 \dots \alpha_n$ and $\beta_0 \dots \beta_n$, then the plant can be written as an LFT as shown in Figure 6.11. where

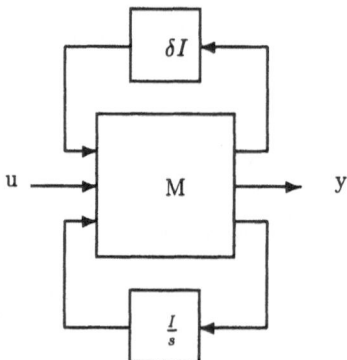

Fig. 6.11. The plant as a linear fractional transformation.

$$
M = \begin{bmatrix}
0 & 0 & I & & & & & & \\
& & & I & & & & & \\
& & & & \ddots & & & & \\
& & & & & \ddots & & & \\
& & & & & & I & & \\
& & & & & & & I & \\
& & & & & & D & C \\
\beta_p & \alpha_p & \dots & \dots & \beta_1 & \alpha_1 & \beta_0 & \alpha_0
\end{bmatrix}
$$

and δ represents the scheduling parameter. It is assumed that δ is scaled such that $-1 \le \delta \le 1$ represents the complete schedule.

Similarly an exact LFT representation of the controller schedule of the form in Figure 6.12 can easily be constructed.

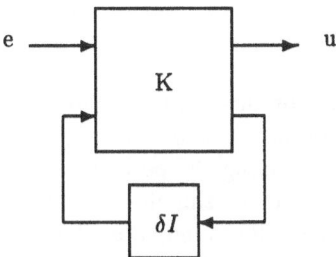

Fig. 6.12. The scheduled controller as a linear fractional transformation.

A schedule on two parameters can also be written in this way, the final LFT having two repeated real blocks. In the single scheduling variable case the plant LFT and controller LFT can now be combined to represent the closed-loop by an LFT with a single repeated real block to represent the schedule as in Figure 6.13.

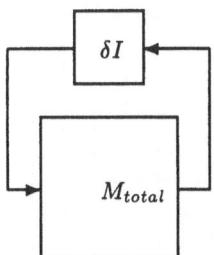

Fig. 6.13. Total LFT representing plant and controller.

Now the schedule is written in this LFT form, the structured singular value (μ) and variations on the structured singular value can be applied. As δ corresponds to speed in our example, it is a real parameter. Assuming that it has been scaled such that $-1 \leq \delta \leq 1$ represents the speed schedule from 0 to V_{max}, then a real structured singular value test (i.e. the fact that δ is real is accounted for) giving $\mu < 1$ indicates that the closed-loop is stable for all frozen operating points on the schedule. The real structured singular value in this case is given by

$$\rho_R(M) = \max\{|\lambda| : \lambda \text{ is a real eigenvalue of } M\}$$

(see [24]), and so can be calculated exactly. However, this does not allow for uncertainty in the LFT model of the plant, nor does it check performance.

To address both of these points an extra complex block can be added to represent the $\mathcal{H}_\infty$/loop-shaping cost function. This is achieved by using the observation in [57] that the coprime factor cost function is the same as:

$$\left\| \begin{matrix} (I - GK)^{-1} & (I - GK)^{-1}G \\ K(I - GK)^{-1} & K(I - GK)^{-1}G \end{matrix} \right\|_\infty$$

Figure 6.14 shows the total M-block for the combined schedule and performance blocks. The corresponding block structure is thus one repeated real block plus one complex block. In this case, assuming the performance block has been suitably scaled by a scalar $1/\alpha$, a structured singular value test giving $\mu < 1$ for mixed repeated real and complex blocks indicates that as well as being stable, robustness and performance are good for all frozen values of the scheduling variable; to be precise, the coprime factor cost, γ, satisfies

$$\gamma \leq \alpha$$

at all points on the schedule.

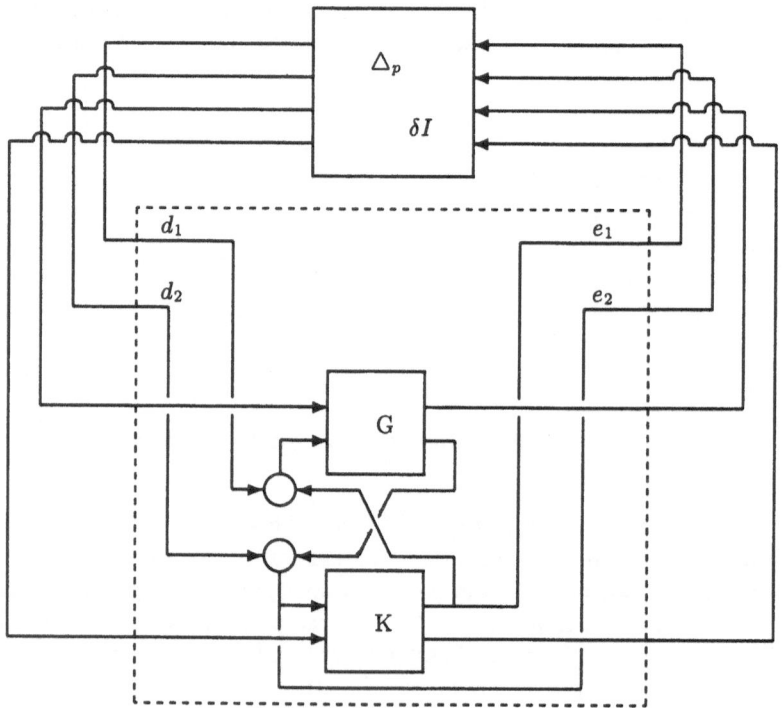

Fig. 6.14. Scheduled system performance test.

In the general case of scheduling on q variables, the total block structure would be one complex block plus q repeated real blocks. At present software is not available to solve μ exactly for repeated real blocks. However, upper bounds can be found by for example treating the repeated blocks as complex. This approach of writing the plant and controller as LFT's so as to use the μ framework offers great potential to analyse gain schedules.

Note that the tests described so far have not allowed the scheduling variable to vary with time. The μ-analysis formulation also offers potential for evaluating the time-varying properties of the schedule by restricting the D's in the DMD^{-1} upper bound for μ to be invariant with frequency. A way of imposing this restriction on the D's is given in [23]. This time-varying test would check for instabilities caused by the schedule. However, it is likely to be very conservative in that the test imposes no restraint on the rate of change of the scheduled variable which in practice there would be.

6.4.2 Application to the forward speed schedule

Figure 6.15 shows $\rho(M)$ (the spectral radius) and $\rho_R(M)$ for the combined controller and plant LFT's where the controller is the two-design schedule. The plant LFT was formed by fitting a 5th order polynomial through the

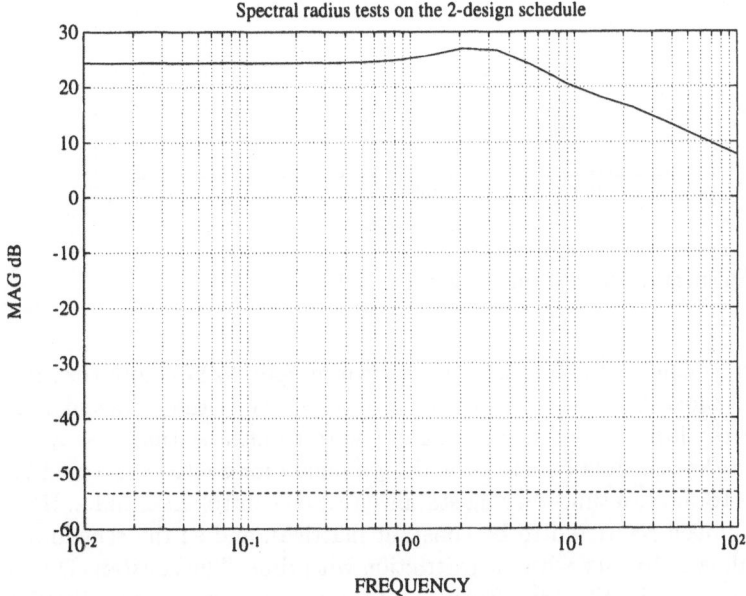

Fig. 6.15. Spectral radius (—) and real spectral radius (- -) for 2 design schedule.

coefficients of 6 plant linearisations in the 0 to 100 knot speed range. The plot shows that for analysis of the schedule, treating δI as complex is too conservative. Remember also that this plot only checks stability for frozen operating points on the schedule i.e. the scheduling variable is assumed non-timevarying. If we are to check performance for each frozen operating point of this schedule by adding a complex block to represent performance, then we need to be able to calculate μ for repeated real blocks combined with a complex block. However, for two adjacent designs from the 6 design schedule, treating δ as complex was not too conservative as can be seen from Figure 6.16.

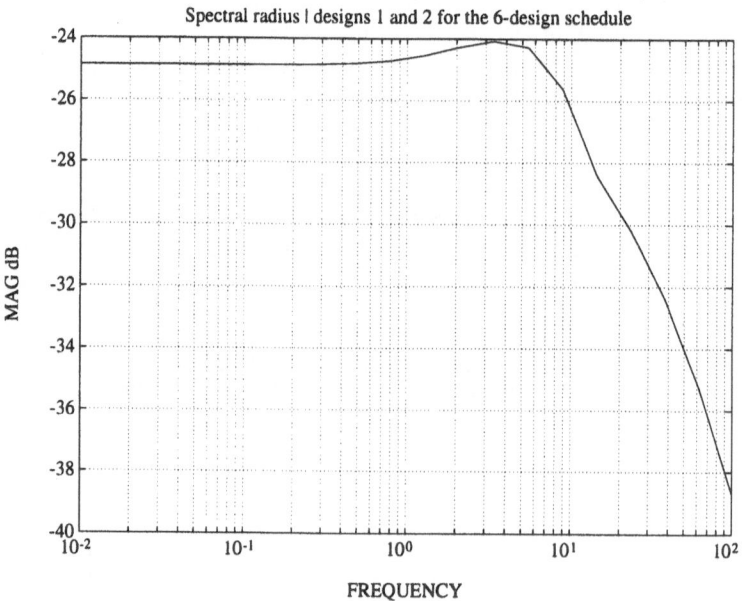

Fig. 6.16. Spectral radius test for designs 1 and 2.

The performance μ for this set-up is shown in figure 6.17 - note that the performance block has been scaled by $1/\gamma_1$ where γ_1 is the achieved cost function for the lower speed linear design. The test indicates that $\gamma \leq \gamma_1$ for all points on the schedule between the design points for controllers 1 and 2.

Figure 6.18 shows a time-varying stability test on controllers 1 and 2. Here the D's have been restricted to be constant matrices, and so the scheduling variable is allowed to vary without restriction with time. The constant D was selected by taking the D at the frequency for which the structured singular value (non-timevarying test) is a maximum ($4 rads^{-1}$), and as such makes the test more conservative than it needs to be. An alternative approach to this ad-hoc scheme to find the optimal D-scaling can be found in [23]. The test is

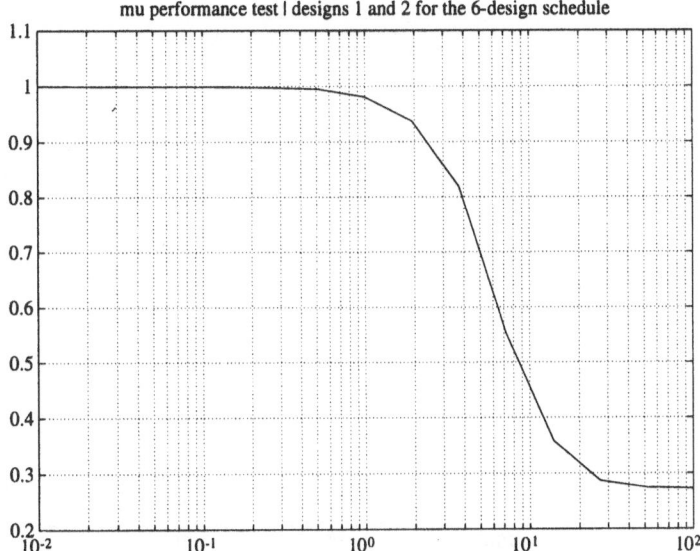

Fig. 6.17. Structured singular value performance test for designs 1 and 2.

too conservative, which is not surprising as in reality the scheduled variable can not vary infinitely fast.

The potential benefits of this type of scheduling test are much greater for schedules on several variables. At the time of writing software capable of handling this complexity for several repeated real blocks was only just becoming available, and so has not as yet been applied to the schedule on both speed and incidence.

6.5 Summary

It has been demonstrated how $\mathcal{H}_\infty$ controllers from the normalised coprime factor/loop-shaping approach of McFarlane and Glover can be scheduled by using its unique observer structure. This combined linear design and scheduling approach has proved to be very versatile when applied to the GVAM, and may well be suitable for many other applications. The whole design procedure provides a systematic way of incorporating knowledge of plant non-linearity captured by key variables into the design. The inherent robustness achieved by the $\mathcal{H}_\infty$ controller can be left to cope with unmodelled dynamics and modelling errors. From the work of Shamma and Athans discussed at the start of this chapter, this robustness means that problems are less likely to be encountered whilst the schedule is traversed. The ability to cope with more than one key (scheduling) parameter is a particularly attractive feature.

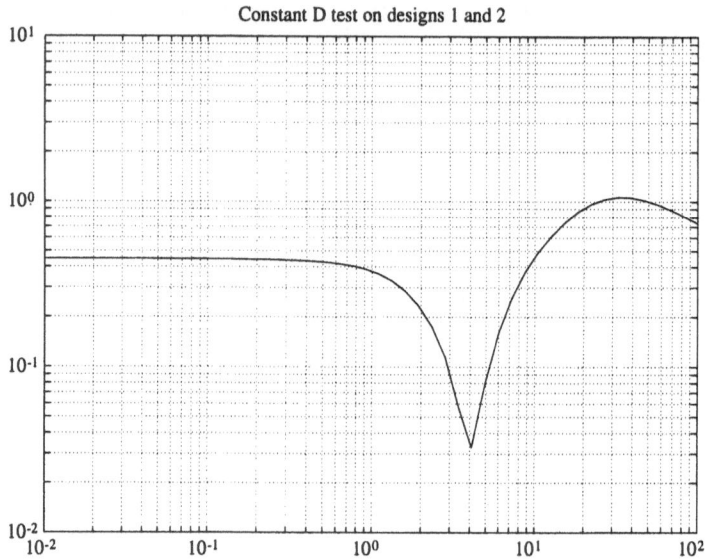

Fig. 6.18. Time-varying test on designs 1 and 2.

With the GVAM it was found that close to optimal performance in terms of linear optimisation cost could be achieved with a modest number of design points. The observer structure has a couple of additional advantages. For example, the structure has a clear physical meaning. This is very important if practising control and aero-engineers are going to accept these type of controllers. There is also more potential with this structure for a high level algorithm to monitor the states of the controller, and check for controller, sensor or actuator failure. The observer structure can also be used to implement anti-windup procedures for saturating actuators, and this possibility is discussed in the next chapter. However, the implementation is rather complex in that a lot of parameters have to be scheduled.

The scheduled system potentially lends itself well to analysis. In particular, the use of the structured singular value offers great potential for checking performance of designs with schedules on several parameters, the advantage being that the number of time-costly time simulations required to evaluate a control system could be reduced. For time-varying analysis it is possible that a less restrictive condition could be placed on the D scalings which is related in some way to the maximum rate of change of the scheduling variable.

CHAPTER 7
MULTIVARIABLE ANTI-WINDUP

7.1 Saturating actuators on multivariable systems

Linear design is typically carried out without regard to what the system behaviour will be should an actuator saturate or rate-limit. The gains of the control loop are selected so that actuators are sufficiently utilised to meet normal closed-loop performance requirements. Selecting the gains to avoid saturating the actuators when large disturbances or commands act on the closed-loop would give poor performance under normal operating conditions.

To see the effect of a saturation on a single-loop proportional plus integral (P+I) control system consider Figure 7.1. The saturation of the plant

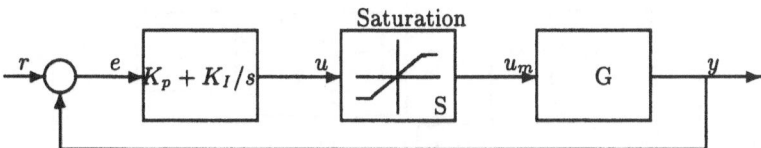

Fig. 7.1. Plant actuator limitations.

actuator is modelled by the saturation block, S. Consider a large step on r which causes u_m to saturate; y cannot track r, and so e becomes large. The integral term $\frac{K_I}{s}$ of the controller integrates e causing u to further increase and send the actuator further into saturation. Eventually y reaches r, and e goes to zero; however u is still high due to the accumulated value of the integrator state of the controller, and y overshoots it's set point. The output y will take some time to settle while the integrator state decays away; this effect is known as integrator wind-up. One well accepted solution to this phenomenon for classical single-input single-output systems is just to freeze the integrator whenever the plant actuator saturates, and this can often be all that is required.

Multivariable systems present much more of a problem when actuators saturate, and anti-windup compensators have to be designed a lot more carefully. This is because the loop-gain now has both magnitude and direction which are both affected by saturating actuators. The loss in directionality can mean loss of decoupling between the controlled outputs. The situation is made much worse if more than one actuator saturates at any one time.

Several schemes exist for the design of multivariable anti-windup compensators. Although they all tend to be rather ad-hoc, by using a combination of these techniques reasonable anti-windup compensators have been developed for GVAM designs presented here. Although a consistent theory is not available for the design of anti-windup compensators, considerable progress has recently been made in [49],[50] to unify these various schemes into one framework. This framework offers potential in the future for formulating the anti-windup compensator design as an optimisation. This framework is summarised in §7.2, and following this four anti-windup compensator design approaches are given in §7.3— §7.6. Finally the key points of the chapter are summarised in §7.7.

7.2 A framework for anti-windup analysis.

In [50] some of the existing ad-hoc anti-windup techniques are analysed with reference to the set-up in Figure 7.2. The classical approach of modifying

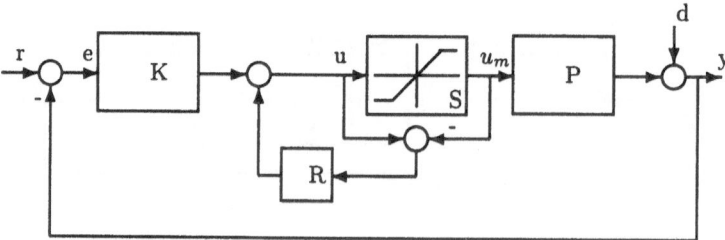

Fig. 7.2. An anti-windup set-up.

error integration during saturation is shown in [50] to be modelled by $R(s) = \frac{1}{\alpha s}I$ for example; putting $R(s) = \frac{1}{\alpha s}I$ into the closed-loop equations during saturation moves the integral terms to poles at $-1/\alpha$. Most of the schemes discussed in this chapter can be written in this form. As has already been described in chapter 5, some progress has been made in [49] for analysis of such schemes by conic sector bounding the saturation non-linearity, and using the time-varying version of the structured singular value. A general theory for multivariable Anti-Windup and Bumpless Transfer (AWBT) is also developed in [49]. Although this theory is not directly used for the anti-windup compensator designs in this thesis, it gives some insight into how the various schemes described later in this chapter relate to each other. Figure 7.3 illustrates the problem formulation for the AWBT theory.

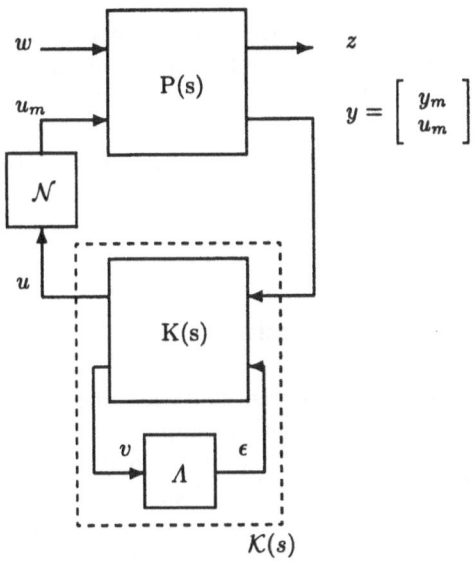

Fig. 7.3. An AWBT framework.

Signals w and z represent the standard disturbance and performance error vectors, $\mathcal{N}$ is the saturation or constraint, and $\mathcal{K}(s)$ the AWBT compensated controller. Λ is the AWBT operator. Notice that the controller has access to plant inputs as well as plant outputs. The following assumptions are made:

1. Λ is causal, linear, and time-invariant.
2. For all t, $u - u_m = 0 \Rightarrow \epsilon = 0$.

Note that as a consequence of the second condition the anti-windup operator Λ must be a constant matrix. Under these assumptions, the result is obtained that the AWBT design is equivalent to selecting a factorisation $K(s) = V^{-1}(s)U(s)$ for the linear controller. The controller is then implemented as

$$\mathcal{K}(s) = \left[\begin{array}{cc} U(s) & I - V(s) \end{array} \right]$$

As a criterion for choosing this factorisation they suggest that performance as the actuator goes into or comes out of saturation is related to controller "memory"; slow modes in the controller make for a sluggish recovery when coming out of saturation. They propose choosing a factorisation to minimise the Hankel norm of $\mathcal{K}(s)$ as this is a measure of how much past disturbances affect future outputs. However there is a trade-off between making the Hankel norm small and susceptibility to noise. This is less of a problem if the saturated plant input levels are software generated as opposed to being actual measurements of the plant inputs.

This method of design does have a lot of appeal, but does have some limitations. The designer does not have the same flexibility that there is with the high gain approach in §7.6 to prioritize the order in which outputs are backed-off when a saturation occurs. The assumptions made in the derivation are also restrictive to some extent. Allowing Λ to be non-linear may have significant advantages, and indeed not all the schemes outlined in the next sections have Λ as linear. The motivation for the second assumption is lost by simply disconnecting the anti-windup compensator when the system comes out of saturation. This would then allow Λ to be dynamic.

7.3 Application of Hanus' approach

The Hanus approach [6] for switching between controllers as described in chapter 5 is equally applicable to controller anti-windup in the event of actuator saturation. All that is required to extend the switching set-up to include desaturation is to drive the on-line Hanus self-conditioned controller with actual (possibly saturated) plant inputs rather than controller outputs. This is illustrated in Figure 7.4.

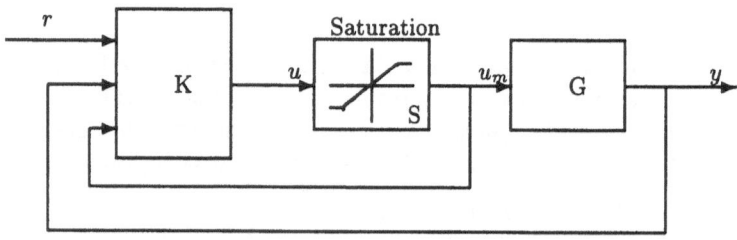

Fig. 7.4. Hanus self-conditioned structure applied to desaturation.

Just as when the technique is used for switching the scheme relies on updating the controller with a set of inputs which would have given present plant inputs. This is essentially the same as running the controller backwards on the saturated plant inputs. The same limitations on the controller in terms of invertibility as were discussed in §5.2.3 apply.

A design study on the GVAM [58] showed that this scheme works well for single actuator saturations. However, it was found that if more than one actuator saturates, then decoupling can degrade significantly. Essentially a large demand or disturbance on one of the plant outputs can cause the directionality of the control vector at the plant input to be lost, and so results in large coupling into other outputs.

7.4 Variable loop-gain approach

The variable loop-gain approach to be described here was developed as a result of the deficiencies of the Hanus approach experienced when applying it to the GVAM [58]. Independently the scheme was also proposed in [59]. The structure of the scheme is shown in Figure 7.5.

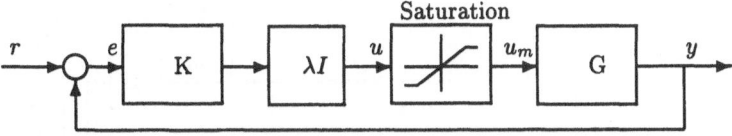

Fig. 7.5. Variable loop-gain desaturation.

Under normal operation $\lambda = 1.0$. If however any one of the elements of the control input vector u_m saturates, then λ is reduced until $u = u_m$ i.e.

$$\lambda = \frac{1}{\max_i(|u_i - u_{mi}|) + 1}$$

The effect of this scheme is to lower the bandwidth of all loops, and to preserve the directionality of the control input vector u_m.

Stability results are given in [59] for when both the plant and controller are open-loop stable; if either the plant or compensator is unstable, then the controlled system will become unstable for small λ. Notice that this scheme does not satisfy the assumptions used in the framework in [50] in that the scheme is non-linear.

7.5 Observer approach

Another anti-windup scheme is to implement the controller with an observer structure. If instead of using controller outputs, the actual plant inputs are used to drive the observer, then the controller states will remain consistent with the plant states. As was discussed in chapter 6 when looking at ways of gain scheduling, the $\mathcal{H}_\infty$ controller from the normalised coprime factor robust stabilisation problem can be written as an exact plant observer.

Figure 7.6 is a block diagram of the plant, weights and the controller in observer form. With the implementation shown, windup will still occur as the observer is driven by u rather than the actual limited plant inputs. If, however, we replace G_W (enclosed in the dotted lines) with the block diagram in figure 7.7, then both W_1 and the feedback controller will be protected against wind-up. To see that the feedback controller is desaturated is easy

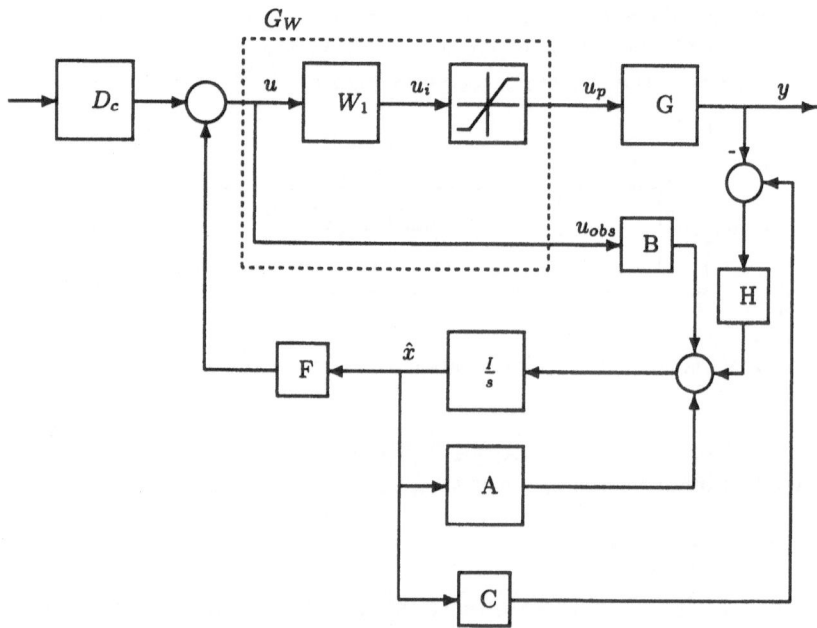

Fig. 7.6. Utilising the observer structure for anti-windup.

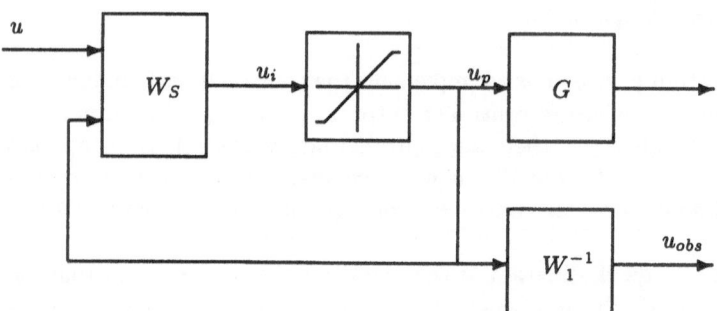

Fig. 7.7. Self-conditioned form of the precompensator.

- it is driven directly from actual plant inputs via W_1^{-1}, which is the same as driving it with $\bar{u}$ where $\bar{u}$ is the value of u which would have plant given input u_p had the system been completely linear. W_1 has been replaced with a Hanus [6] self-conditioned form, W_S. If the state-space realisation of W_1 is given by

$$
W_1 = \left[\begin{array}{c|c} A_w & B_w \\ \hline C_w & D_w \end{array} \right]
$$

then

$$
W_S = \left[\begin{array}{c|cc} A_w - B_w D_w^{-1} C_w & 0 & B_w D_w^{-1} \\ \hline C_w & D_w & 0 \end{array} \right]
$$

It is easy to verify that in the event of no saturation that the resulting transfer function from u to u_p is W_1. In the event of saturation, the implementation of W_S is equivalent to running W_1 backwards on u_p i.e. the states will always be consistent with current plant inputs. Hence the desaturation action. Control law 005 also has an align matrix, M. Taking this into account, the overall scheme can be implemented as:

$$
\left[\begin{array}{c} u_i \\ u_{obs} \end{array} \right] = \left[\begin{array}{c|cc} A_w - B_w D_w^{-1} C_w & 0 & B_w D_w^{-1} \\ \hline C_w & D_w M & 0 \\ -M^{-1} D_w^{-1} C_w & 0 & M^{-1} D_w^{-1} \end{array} \right] \left[\begin{array}{c} u \\ u_p \end{array} \right]
$$

Note that both M and M^{-1} are required for the implementation, but the desaturation scheme does not increase the degree of the weighting function.

7.6 Prioritised approach

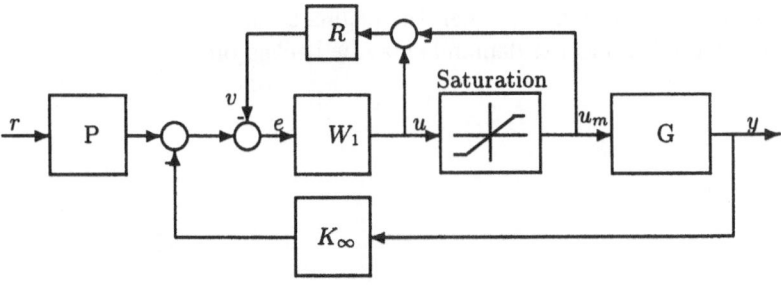

Fig. 7.8. Prioritised desaturation approach.

The prioritized approach is described with application to the GVAM in [4]. Figure 7.8 shows the set-up. Remember that W_1 contains any integral action put in by the designer. In the event of saturation the error signal is backed off by v so that the states of W_1 do not wind up. Typically $R = kI$ where k is a large constant. This fits into the framework in [50] where R is assumed to be a constant matrix. The larger k, the less the states of W_1 wind up. However, there is usually a limit to how high k can be as the anti-windup loop will go unstable. This suggests that in some cases it may be desirable to put some phase compensation in R to enable the gains to be pushed higher. Putting $R = kI$ assumes that the plant inputs and outputs are ordered sensibly; a saturation on the first input will back off tracking performance on the first output. The great power of this approach is that the designer can decide which outputs/references to effectively back-off when a particular actuator saturates by making R a function of flight condition and demand settings.

7.7 Summary

A set of anti-windup techniques have been discussed. Although they are ad-hoc in nature, it will be shown in chapter 9 that between them they are able to give good anti-windup protection for the GVAM designs. Successful anti-windup is central to the implementation of multivariable controllers. The framework in [50] has been outlined, and it has been discussed how the ad-hoc schemes relate to this. This framework offers the most consistent approach to anti-windup available to date, but as has been discussed relies on assumptions which are broken by some very computationally simple ad-hoc techniques.

The observer approach presented here has altered somewhat from what is presented in [4]. The original formulation did not include the combined use of the Hanus and observer approaches. It provides a very neat and reliable way of effecting desaturation, and does not require any time-consuming and ad-hoc setting up of gains as with the prioritized approach. Its only drawback is that the designer can not influence which demand should be backed off in a particular situation e.g. when the engine saturates in transition, either forward of vertical motion demands may be backed off.

CHAPTER 8
CONTROLLER COMPLEXITY

8.1 Introduction

For implementation of controllers there is a strong drive for simplicity. Simplicity results in reduced hardware, which in turn means a cheaper control system which weighs less and takes up less space. Simplicity may also result in greater integrity of the control system which is vital for aerospace applications. The penalty for large hardware requirements is compounded by the required duplication of controller implementation – for example the A320 Airbus triplicates all hardware for the control law, plus there is a fourth independently designed back-up control law [60]. In Part II, the implementation of control law 005 is restricted by existing hardware in the aircraft, and complexity becomes a central issue.

Often $\mathcal{H}_\infty$ and other optimal control design techniques are labelled as producing unnecessarily complex control laws. Considerable improvements have been made since the first solution methods for $\mathcal{H}_\infty$ appeared, the state space solution methods of Glover and Doyle [15] now producing controllers of order no higher than that of the weighted plant. Model reduction techniques can also be applied to the initial plant model or the final controller. However, it is not just the order of the controller which is important. Structure is also highly relevant, and the apparent lack of structure with $\mathcal{H}_\infty$ controllers can be a serious disadvantage when compared with the simplicity of the structure of a conventional P+I controller.

Complexity depends on the number of calculations required each frame time, and also on the number of controller parameters which must be stored. The number of states is not always a good indicator of complexity; through a different state-space realisation it is often possible to significantly reduce the number of required calculations. This is clearly illustrated by the modal form which will be described in §8.2, and which was used to implement the switched designs of chapter 5 on the RAE simulator in [4].

The necessity for switching/scheduling with operating point also impacts on the resulting implementation complexity. It results in higher computation requirements plus more parameters to store. The relative complexities of the switching and scheduling approaches are compared in §8.3.

The complexity of the scheduling approach largely arises form the large number of parameters in the observer structure used to implement the schedule. The observer structure is used because of the necessity of having a fixed physical structure from one controller design to the next. This motivated a design study using the Multi Objective Programming System (MOPS) at Deutsche Forschungsanstalt für Luft und Raumfahrt (DLR) [61]. MOPS enables the controller structure to be fixed, and its parameters optimised. The results of this study are given in §8.4, and they give some useful insights into the GVAM $\mathcal{H}_\infty$ designs and the trade-off between complexity and performance. Finally the main points and conclusions of the chapter are given in §8.5.

8.2 Implementation using a discrete modal form

The number of calculations required to update a state-space realisation of a controller can be considerably reduced using a so-called modal realisation. Nett and Polley proposed a modal canonical form in [62] for implementation of controllers. When implemented in this form the A-matrix has elements only on the leading diagonal, and the B-matrix is mostly ones and zeros. The realisation is arrived at by taking canonical factorisations of the residues corresponding to the poles of the system. This approach is well suited to controllers designed by such approaches as the characteristic locus method where the designer directly chooses the controller poles, and typically they are chosen to be real. Controllers designed by optimal control techniques are likely to have complex poles, and hence some modification is required as the A-matrix cannot be diagonal and real.

The approach taken for implementation of switched controllers in chapter 5 was to use a modal realisation where the A-matrix only has elements on the leading diagonal and the immediate off-diagonals. Finding such a realisation is very straightforward, and just requires an eigenvalue decomposition. If the initial realisation is $\{A, B, C, D\}$ and $W = [w_1 \ldots w_n]$ is the matrix of eigenvectors of A, then the modal form is given by $\{T^{-1}AT, T^{-1}B, CT, D\}$ where

$$T = [t_1 \ldots t_n]$$

and

$t_i = w_i$ if λ_i is real, else

$[t_i \ t_{i+1}] = [Re(w_i) \ Im(w_i)]$ if λ_i is complex.

Note that modification to this might be required if A has repeated poles as this could make T non-invertible.

For the modal implementation of an 18-state Hanus self-conditioned controller for the GVAM the computation savings are now estimated. Remember that the controller is of the form:

$$
u = \left[\begin{array}{c|c} A & B \\ \hline C & D \end{array} \right] \left[\begin{array}{c} r \\ y \\ u_m \end{array} \right]
$$

where $A \in \mathcal{R}^{18 \times 18}$, $B \in \mathcal{R}^{18 \times 9}$, $C \in \mathcal{R}^{3 \times 18}$, and $D \in \mathcal{R}^{3 \times 9}$. Without using a modal form, the total number of parameters which need to be stored is $18 \times 18 + 18 \times 9 + 3 \times 18 + 3 \times 9 = 567$. Updating the state-space requires $18 \times (18 + 9 - 1) = 468$ additions and $18 \times (18 + 9) = 486$ multiplications, and updating the output equation requires $3 \times (18 + 9 - 1) = 78$ additions and $3 \times (18 + 9) = 81$ multiplications. Hence the total is 546 additions and 567 multiplications.

If a modal form is used, then the state equation requires $18 \times (3 + 9 - 1) = 198$ additions and $18 \times (3 + 9) = 216$ multiplications, and the output equation requires 78 additions and 81 multiplications as before. Hence the totals are 276 additions and 297 multiplications. Note that this is an upper estimate as some of the poles of A may be real, thus giving zero off-diagonal elements for A.

It can be seen that the number of operations required to update the controller is roughly halved using the modal form. Also the number of parameters is which need to be stored is reduced to $3 \times 18 + 18 \times 9 + 3 \times 18 + 3 \times 9 = 297$. Hence both processor computational and memory requirements are greatly reduced. Once diagonal, the controller is easily discretised using the Tustin bilinear transform, the diagonal structure being preserved.

8.3 Complexity of the switching and scheduling approaches

Both the switched and scheduled designs of chapters 5 and 6 are computationally expensive, and require considerably more memory to store parameters than other control laws implemented on the R.A.E. simulator in the past. The complexity is not just due to the relatively high order of the implemented controllers. The necessity for switching or gain scheduling adds to the complexity: switching requires that an off-line controller is updated as well as the on-line controller, whereas scheduling requires interpolation of controller parameters before the states can be updated.

The computational requirements for a Hanus self-conditioned controller has already been calculated in §8.2. The off-line controller only needs the state equation updated giving an extra 198 additions and 216 multiplications assuming a modal implementation. This gives a total of 474 additions and 513 multiplications.

The number of additions and multiplications for the scheduled observer form is now evaluated. Recall the observer equations:

$$
\dot{\hat{x}} = A\hat{x} + H(C\hat{x} - y) + Bu
$$

$$u = F\hat{x}$$

where $A \in \mathcal{R}^{18 \times 18}$, $B \in \mathcal{R}^{18 \times 3}$, $C \in \mathcal{R}^{3 \times 18}$, $F \in \mathcal{R}^{3 \times 18}$, and $H \in \mathcal{R}^{18 \times 3}$. The C-matrix is all zeros apart from three '1' entries as measured outputs are three of the states. The A-matrix is assumed not to have any zero entries. In practice it does, but they are not located in a sufficiently orderly manner for advantage to be easily taken of it in the way it is for a modal A-matrix.

For the state equation total additions and subtractions is $18 \times (18 + 3 + 1 + 3 - 1) = 432$, and the number of multiplications is $18 \times (18 + 3 + 3) = 432$. The output equation requires $3 \times (18 - 1) = 51$ additions and $3 \times 18 = 54$ multiplications. In addition to this, each of the $(18 \times 18 + 18 \times 3 + 18 \times 3 + 3 \times 18) = 486$ parameters must be calculated by interpolation of the form:

$$a = \lambda a_0 + (1 - \lambda)a_1$$

For each update of the controller λ is fixed so $(1 - \lambda)$ only need be calculated once for all parameter updates. Hence each parameter update requires one addition and two multiplications. Hence total computation for the controller each cycle is $432 + 51 + 486 = 969$ additions and $432 + 54 + 486 \times 2 = 1458$ multiplications. Table 8.1 summarises the relative complexities of the switched and scheduled designs.

Table 8.1 – control law computational requirements			
Design	Additions	Multiplications	Parameters per design
Switched	474	513	297
Scheduled	969	1458	486

Clearly the scheduled form is heavier computationally. However, it would be possible to update the controller parameters every other cycle, and thus reduce the amount of interpolation carried out. There is also clearly potential for parallel processing.

Comparison of the number of parameters which must be stored is a little harder. It was shown in [63] that for a comparable upper limit on the achieved coprime factor robustness measure γ across the flight envelope, the two approaches require approximately the same number of parameters when appled to the GVAM. Irrespective of the relative merits of the two approaches, both of them are demanding on computation and memory requirements. In the next section the trade-off between complexity and achieved performance is investigated.

8.4 A design study using fixed structure controllers

8.4.1 A Multi Objective Programming System (MOPS)

A Multi Objective Programming System (MOPS) has been developed by the Institut für Dynamik der Flugsysteme at DLR [64],[65]. The MOPS design procedure minimises a performance index vector over the parameter values of a fixed structure controller. Because the structure of the controller is fixed, gain scheduling of the designs may be possible. Care may be needed to ensure that the optimisation produces continuous parameter gains; starting a design off with the parameters of the adjacent design, and placing design points sufficiently close should ensure this in practice. The MOPS approach also allows the designer some insight into the trade-off between controller structure and achievable closed-loop performance.

The performance index vector can contain requirements on the closed-loop pole locations, closed-loop frequency responses, and closed-loop time domain properties. The performance objectives are carefully formulated to be smooth so that optimisation routines can be applied.

The MOPS was applied to the $0 - 140$ knots part of the GVAM flight envelope [61] at DLR. Linearisations of the GVAM at 6, 86, 122 and 140 knots with the same inputs and outputs as for the 6 knot example in chapter 4 were used. The design aim was primarily to see what performance and stability robustness could be achieved with a simple controller structure, and so initial design specifications were kept to a minimum. The limits on acceptable time-domain performance were taken to be 10% overshoot and 10% coupling for step demands on all three controlled outputs. A time of 1 to 2 seconds to reach 90% of final value following a step demand was also a requirement drawn from the MIL-specifications reviewed in chapter 3.

The first design attempts using MOPS tried to stabilise and give good performance properties for all four linearisations with a fixed gain controller. Then just the 86 knot plant was selected to see what performance was possible at a single operating point, and thus asses the necessity for gain scheduling. The next section summarises the results of the designs.

8.4.2 Designs using MOPS

Design 1 – PID, 4 plants

The first attempt used a PID controller of the form

$$K(s) = K_P + \frac{K_I}{s} + \frac{K_D s}{1 + \epsilon s}$$

where K_P, K_I, and K_D are full 3×3 constant matrices. It was possible to stabilise all four plants with the one compensator, but the time responses were poor.

Design 2 — diagonal lead-lag, 4 plants

As a starting point the dynamic part, P, of the H_∞ design precompensator ,W, was used to shape the plant. A diagonal lead-lag compensator, C, was then cascaded in series with this in the forward loop:

$$
K_{total} = PC = \begin{bmatrix} \frac{s+3}{s} & & \\ & \frac{s+3}{s} & \\ & & \frac{(s+3)(s+2)}{s(s+20)} \end{bmatrix} \begin{bmatrix} \frac{a(s+b)}{s+c} & & \\ & \frac{d(s+e)}{s+f} & \\ & & \frac{g(s+h)}{s+i} \end{bmatrix}
$$

The optimisation proceeded thus:

1. Cost function : minimise $J_S = \exp(\max_i Re(p_i))$ where p_i are the closed-loop poles. Initially the Pattern search (PS) was used, giving $J_S = 3.7$. This was followed by a Sequential Quadratic Programming search (SQPEX) search which resulted in a stable system with $J_S = 0.5$.
2. Add a damping requirement, J_D, for poles outside of the circle radius 0.01 centered on the origin: the achieved values were $J_S = 0.6$, and $J_D = 0.75$. The step responses were poor having large amounts of overshoot.
3. Add step response criteria: it proved impossible to reduce the overshoot

Design 3 — non-diagonal lead-lag, 4 plants

The structure of C was generalised to:

$$
C = \left[\begin{array}{ccc|ccc} ac11 & & & 1 & & \\ & ac22 & & & 1 & \\ & & ac33 & & & 1 \\ \hline cc11 & cc12 & cc13 & dc11 & dc12 & dc13 \\ cc21 & cc22 & cc23 & dc21 & dc22 & dc23 \\ cc31 & cc32 & cc33 & dc31 & dc32 & dc33 \end{array} \right]
$$

thus allowing coupling between the loops, and resulting in 21 parameters to optimise. This gave C the freedom to take on the role of the 'align' matrix used for the H_∞ designs. The cost function objectives were the same as for design 2(3). Only the Powell optimisation caused any change in the parameters, and even the effect of this was minimal.

Design 4 — diagonal lead-lag in the feedback path, 4 plants

Design 2(2) was evaluated with C in the feedback path. The motivation for this comes from the implementation structure used for the H_∞ designs where K_∞ is placed in the feedback path. Overshoot requirements were met in all cases, but rise-times for forward and vertical speed demands were too large, and interaction levels for pitch step responses too high. The pitch response also had an undesirable kink as it reaches about 70% of final demand. Allowing C to have the structure used in design 3 did not improve things.

Design 5 — design for a single operating point

An investigation was made as to what performance is achievable at a single operating point using the 86 knot linearisation. The compensator C was placed in the feedback loop, and the following structures were used:

1. The structure used in design 3 was used as an initial try, starting with the controller parameter values from that design. The cost functions used were stability, damping, overshoot, interaction, and rise-time. This was with the exception of the pitch channel where the overshoot requirement was replaced by an integral of the square error (SQC) cost. The resulting controller still exhibited poor characteristics on the pitch channel.

2. A full lead-lag structure was tried with no improvement i.e. C consisted of 9 first order lead-lag terms.

3. The poles of C were allowed to become complex by using the structure

$$C = \begin{bmatrix} \frac{g_1(s+d_1)(s^2+e_1 s+f_1)}{(s+a_1)(s^2+b_1 s+c_1)} & & \\ & \frac{g_2(s+d_2)(s^2+e_2 s+f_2)}{(s+a_2)(s^2+b_2 s+c_2)} & \\ & & \frac{g_3(s+d_3)(s^2+e_3 s+f_3)}{(s+a_3)(s^2+b_3 s+c_3)} \end{bmatrix}$$

This formulation proved very sensitive, and it was not possible to find a solution which stabilised the system.

4. The structure and cost functions of design 5(1) were used, plus the parameters of the precompensator were allowed to vary. Hence P is now:

$$P = \begin{bmatrix} \frac{g_1(s+n_1)}{s} & & \\ & \frac{g_2(s+n_2)}{s} & \\ & & \frac{(g_3(s+n_3)(s+m_3))}{s(s+d_3)} \end{bmatrix}$$

thus giving a total of 29 parameters to optimise. Starting with the parameter values resulting from design 5(1), interaction was decreased, rise-times improved, and the pitch channel response improved. The step responses are shown in figures 8.1, 8.2 and 8.3.

Design 6 — parameterised P and C, 4 plants

Starting with the controller structure and parameter values of design 5(4), the other three plant models were added to the cost functions to see if a better fixed design for all plants could be found by allowing the optimisation to alter the precompensator P. The optimisation now has 52 cost function objectives to minimise or keep within constraints, and 29 parameters to vary. Five iterations of Powell produced good results (figures 8.4, 8.5, and 8.6), in particular the rise times being much faster than for design 4. Following this 5 iterations of SQPEX produced a slight improvement.

8.4.3 A comparison of H_∞ and MOPS designs

Controller frequency response

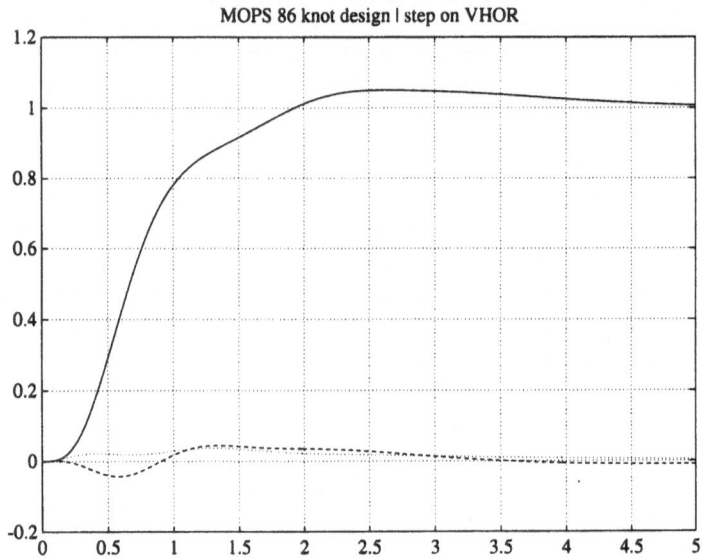

Fig. 8.1. MOPS design 5 — 86 knot operating point design — VHOR step

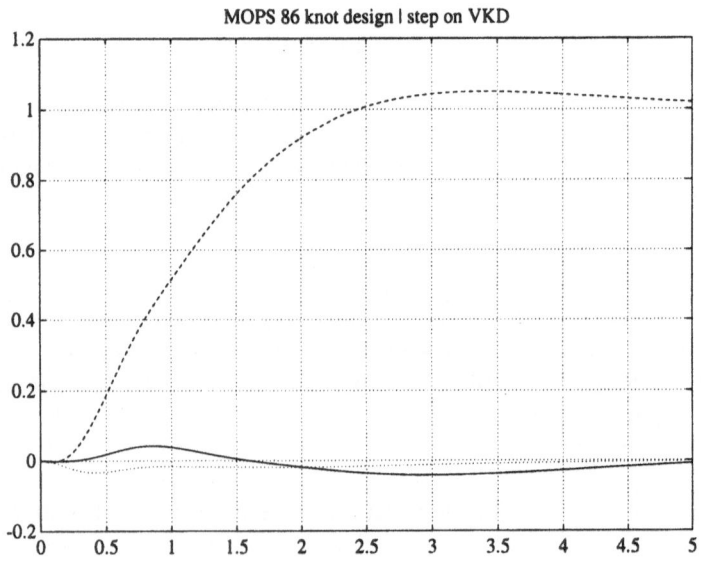

Fig. 8.2. MOPS design 5 — 86 knot operating point design — VKD step

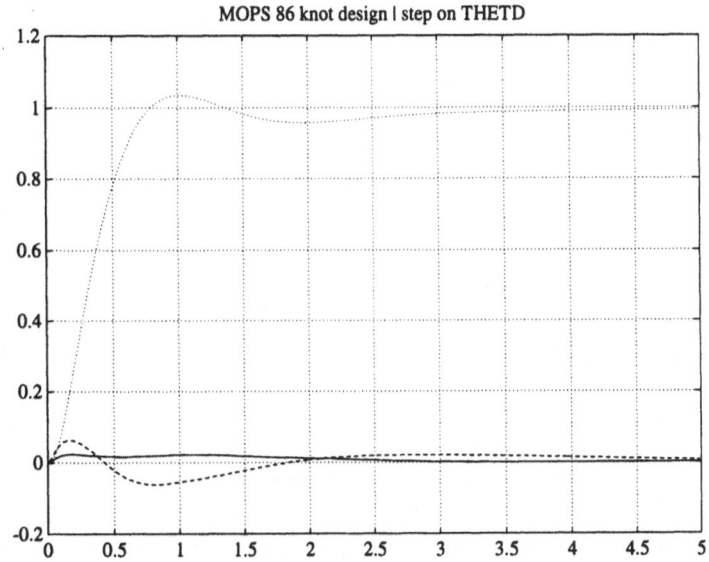

Fig. 8.3. MOPS design 5 — 86 knot operating point design — THETD step

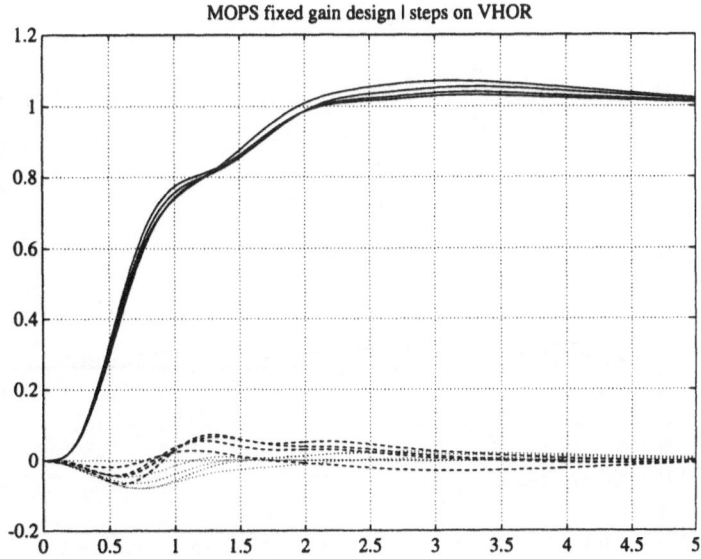

Fig. 8.4. MOPS design 6 — Fixed gain design applied to 6,86,122,140 knot plants

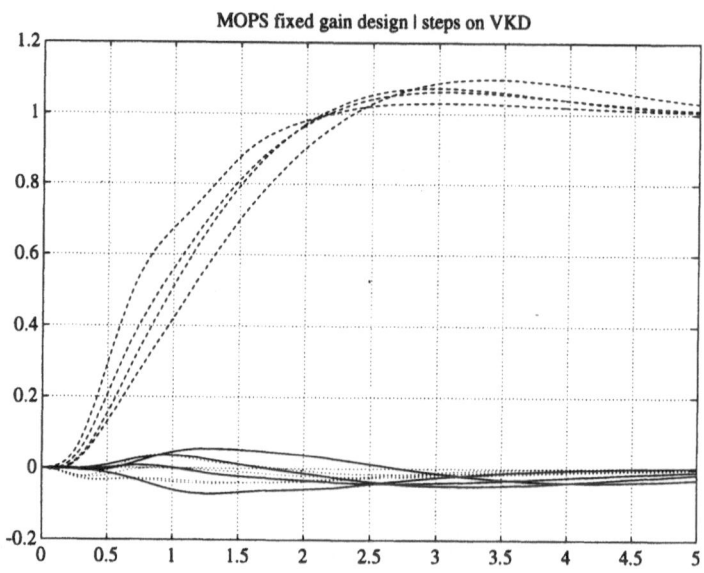

Fig. 8.5. MOPS design 6 — Fixed gain design applied to 6,86,122,140 knot plants

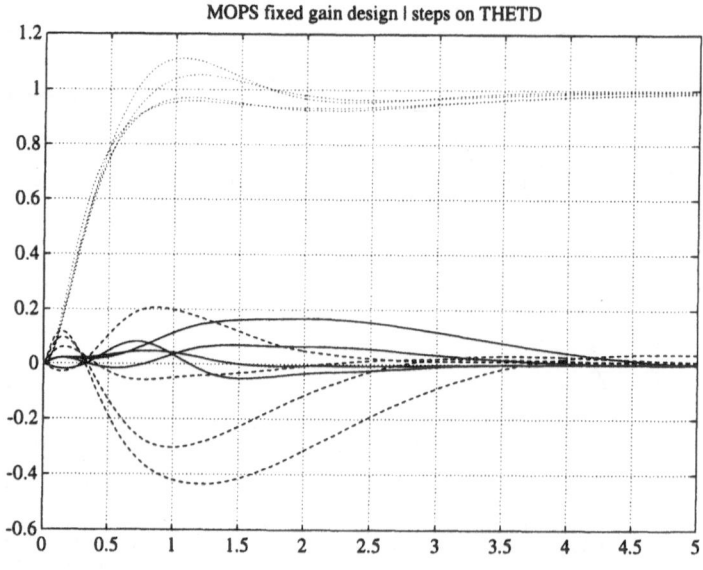

Fig. 8.6. MOPS design 6 — Fixed gain design applied to 6,86,122,140 knot plants

Figure 8.7 shows the frequency response of the (1,1) element of the H_∞ and MOPS controllers for the single operating point overlaid. This is typical of the relationship between the designs for the other elements. The corresponding elements havè similar gains, but the H_∞ controller has extra dynamics at higher frequencies. The off-diagonal elements differ more significantly. The extra dynamics on these elements are likely to be giving rise to the better decoupling observed with the full order $\mathcal{H}_\infty$ design.

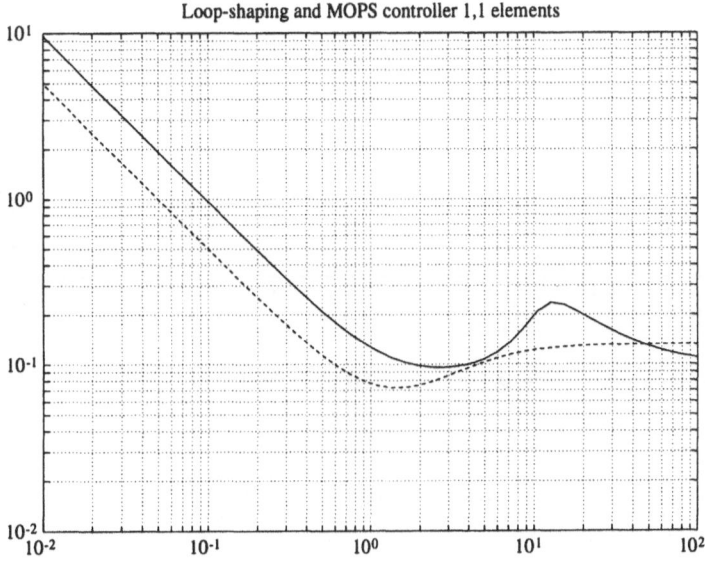

Fig. 8.7. The (1,1)-elements of the $\mathcal{H}_\infty$ and MOPS controllers including weighting functions

Time responses

Figure 8.8 shows the VHOR step response for an H_∞ controller designed for the 86 knot point. Steps on VKD and THETD are similar. It is clearly better than that obtained from the MOPS using simple structure controllers. However, the H_∞ design responses may be better than is actually required; the responses for the fixed operating point MOPS design look acceptable.

The fixed-gain designs attempting to give good performance for all plants exhibit more coupling than is desirable, but certainly have potential as back-up controllers. The H_∞ 86 knot controller exhibits a similar degree of robust performance as can be seen from Figure 8.9. The VHOR and VKD steps exhibit considerably more robust performance than THETD steps.

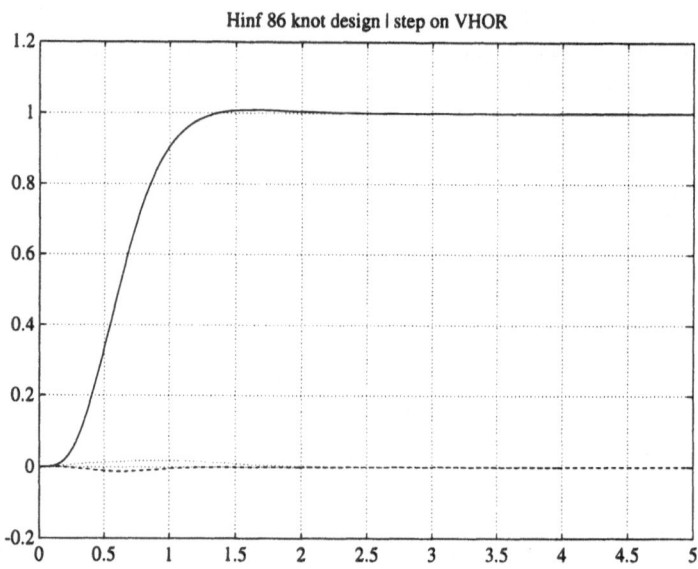

Fig. 8.8. 86 knot $\mathcal{H}_\infty$ design step on VHOR at 86 knots

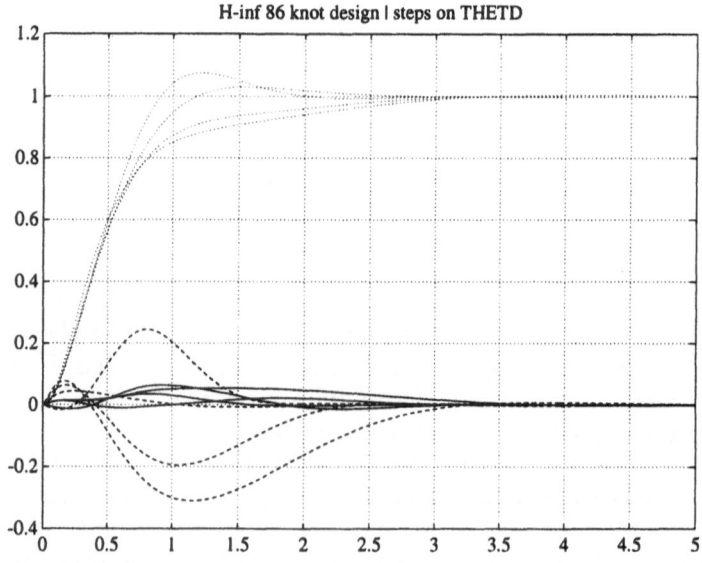

Fig. 8.9. 86 knot $\mathcal{H}_\infty$ design — steps at 6,86,122 and 140 knots

If balanced truncation is applied to the H_∞ controller to reduce the number of states to a comparable level to the MOPS designs, it is not possible possible to obtain as good responses. Figures 8.10 and 8.11 illustrate steps on forward speed for K_∞ reduced using balanced truncation to 8 and 3 states respectively from the initial 18 states. The bounds on the achieved $\mathcal{H}_\infty$ cost function given in [21],[57] which make use of the sum of the tail of the singular values removed were found to be much too conservative for these levels of model reduction.

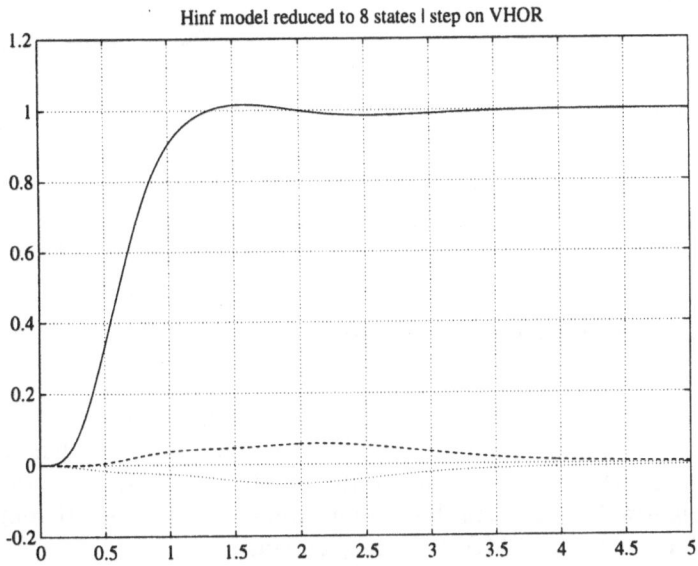

Fig. 8.10. 86 knot $\mathcal{H}_\infty$ design — K_∞ reduced to 8 states

Complexity

The MOPS designs have a clear advantage when it comes to implementation in that only 29 parameters are required, whereas the H_∞ controller has about 500 parameters to store when implemented in observer form. Many of these parameters do not need scheduling, but about half do. The high number of parameters partly arises from the requirement of having a fixed structure (the full observer structure) in order to be able to gain schedule.

Robustness

Comparison of robustness levels between designs is not easy since the choice of the robustness measure used will affect the outcome. Here the robustness to additive perturbations to the normalised coprime factorisation of

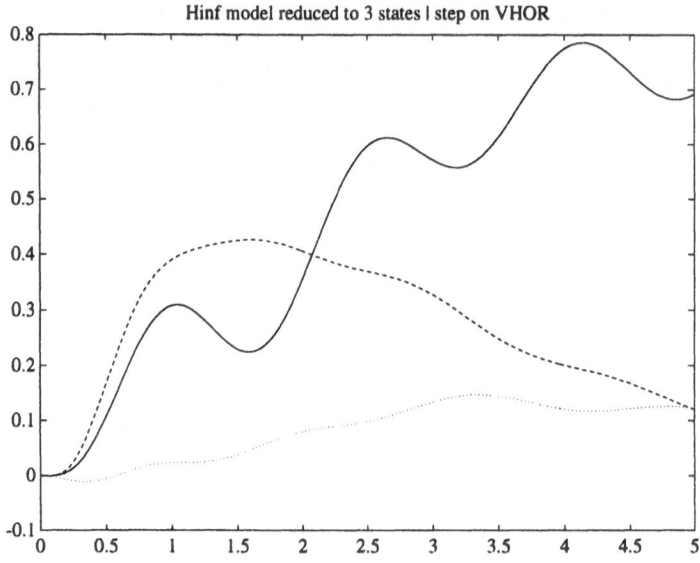

Fig. 8.11. 86 knot $\mathcal{H}_\infty$ design — K_∞ reduced to 3 states

the unshaped plant is used. The unshaped plant is used because the weighting functions for the $\mathcal{H}_\infty$ and MOPS designs are different – note that elsewhere in this thesis the robustness margin referred to ($\epsilon = 1/\gamma$) is for the shaped plant. Figures 8.12 and 8.13 show the maximum size for perturbations to $\tilde{M}$ and $\tilde{N}$ for the $\mathcal{H}_\infty$ and MOPS 86 knot designs.

It can be seen that the full $\mathcal{H}_\infty$/loop-shaping controller (solid line) is more robust at frequencies below the bandwidth ($4 rads^{-1}$), and at higher frequencies there is very little difference between the designs. Essentially both designs have comparable robustness levels, and the extra complexity of the full $\mathcal{H}_\infty$/loop-shaping controller is not gaining anything significant when it comes to robustness as measured in this sense.

Design ease

The H_∞ design approach is very easy to use; how to alter the design weights in order to achieve desired performance is highly intuitive. Also turnaround times from altering a weight to having found the optimum controller and calculated time and frequency responses is typically less than a minute for the GVAM designs using a Sun SPARC SLC with MATLAB and the Mutools toolbox [13], and so the designer can quickly discover what is achievable. The MOPS approach is computationally highly intensive during design, particularly when the number of objectives and parameters becomes large. It may also get stuck at local minima; some experimentation with different optimisa-

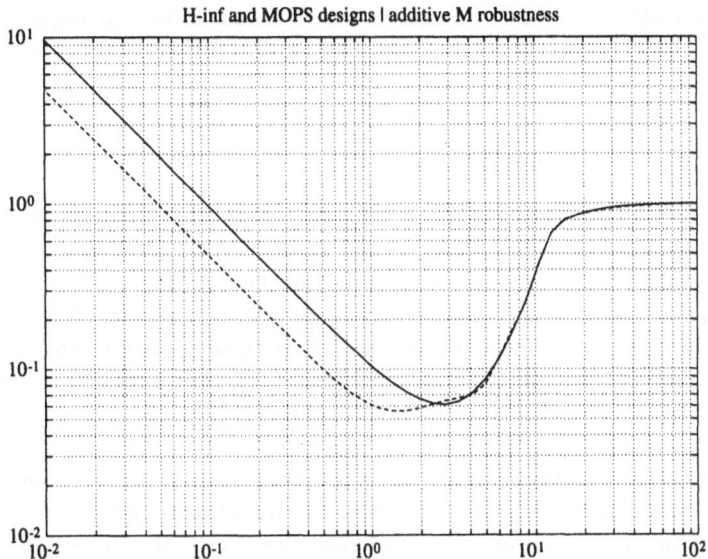

Fig. 8.12. Maximum additive perturbation to $\tilde{M}$

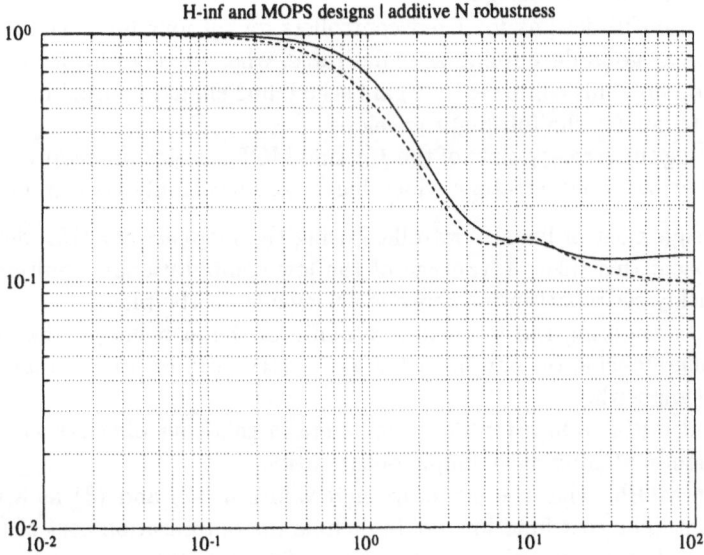

Fig. 8.13. Maximum additive perturbation to $\tilde{N}$

tion procedures was required for the GVAM design examples. However, it is highly versatile in the way objectives may be entered into the design as time-response criteria, frequency domain criteria, or pole location requirements in the complex plane.

8.4.4 Conclusions from the MOPS design study

The MOPS designs clearly indicated that although the H_∞ can give very good control designs in terms of performance and robust stability, the approach does not allow the trade-off between performance and controller complexity to be made easily. This is particularly a problem if gain scheduling is to be used in that model reduction techniques cannot be applied to the controller.

The impossibility of finding a simple controller structure which would give good performance when implemented in the forward loop confirms what was suggested by the H_∞ designs: closed-loop stability requirements conflict with time-domain step response requirements. This suggests that for best robust stability margins the closed-loop controller should be designed with stability cost functions. Time-domain requirements can then be addressed through how the controller is implemented (e.g. in the feedback loop), and/or demand precompensation.

The MOPS designs were improved when the parameters of the precompensator were allowed to vary. This suggests that the precompensator chosen for the H_∞ design places some undesirable restrictions on the time-domain response of the final design.

The MOPS designs for fixed controllers for all four flight conditions do not have as good decoupling as is desirable, whereas those for a single flight condition do. This confirms that gain scheduling should be used for design of a control law for the full flight envelope.

Taking the strong points of the H_∞ and MOPS design procedures suggests the following design strategy which has been successfully tried in [66]:

1. Design a set of linear controllers using the loop-shaping/H_∞ design approach: this takes advantage of the low computational requirements to quickly see what performance is achievable for the plant.
2. By considering the frequency response of the elements of the H_∞ controllers decide on one fixed low order structure to approximate all the designs with.
3. Find initial values for the parameters in this controller structure using model-matching with simple optimisation.
4. Use MOPS with the structure and values in (2) and (3) as a starting point. The cost function would include the optimisation cost of (1), plus possibly time-domain requirements to fine-tune the controller.

Using this approach the MOPS is initiated close to the final design, and computational requirements should be greatly reduced. Potential local minima are likely to pose less of a problem. The final controller would have low complexity, and would be suitable for gain scheduling.

8.5 Summary

The importance of low complexity controllers has been discussed. The computational and memory savings achieved using modal forms has been quantified for the GVAM designs. A comparison of the switched and scheduled control laws for the GVAM indicated that the switched approach is less demanding in terms of processor power. Both control laws are considerably more complex than existing single loop classically designed control laws. However the MOPS design study indicated that this extra complexity does give rise to superior performance, particularly with respect to decoupling.

A weakness of the $\mathcal{H}_\infty$/loop-shaping approach as it stands is that the trade-off between controller complexity and performance is not easily made; in particular model reduction to very few states (from 18 down to 3 states for comparison with the MOPS design in the example here) using balanced truncation gives poorer time-domain performance than can be achieved using parameter optimisation. With balanced truncation of K_∞ it was only possible to reduce the states from 18 to 12 whilst preserving good time-domain properties. Similar limitations were found using balanced reduction of a normalised coprime factorisation of K_∞. Combining the $\mathcal{H}_\infty$ approach and the MOPS approach into one design procedure as proposed in §8.4.4 offers great potential for applications where complexity must be kept low. This scheme has been successfully carried out in [66].

PART II
A FLIGHT CONTROL LAW FOR THE DRA XW175 RESEARCH HARRIER

CHAPTER 9
BACKGROUND

The DRA research Harrier XW175 [67] is a converted T4 Harrier trainer. The front pilot acts as a safety pilot, and the cockpit controls have their conventional functions. The rear pilot acts as a test pilot, and the rear cockpit controls do not have their conventional functionality. Instead they act as demands to an experimental flight control law which in turn sets the nozzle, throttle, tailplane and flap servos. The safety pilot is able to take control from the control law at any time by opposing the motion of the cockpit inceptors which are back-driven from the servos. By restricting the flight envelope such that the safety pilot is always able to regain control, new flight control laws can be very quickly taken into flight.

Flight Control Law 005 is described in this second part of the monograph. It utilises the techniques developed and tested in Part I, and is designed for both piloted simulation studies on the DRA simulator and flight testing on the DRA research Harrier. The control law differs significantly from that presented in [4] in that it uses gain scheduling (as opposed to switching) and uses a different feedback structure which is more suited to implementation on the aircraft.

The control law is for the longitudinal motion for speeds between hover and 250 knots as this is the current flight envelope for the XW175 VAAC research Harrier. It is aimed primarily at the task of recovery to an aircraft carrier which currently places a high workload on the pilot. As the main aim is reduction in pilot workload, a two inceptor demand system has been used. In doing this a trade-off between reducing operational flexibility and reducing pilot workload has had to be made. However, the operational envelope is likely to be widened (for example, extended poor weather operability). It is envisaged that if such a control law were used in production aircraft, it would only be active for recovery to the carrier and for take-off and transition to wing-borne flight - it could perhaps become active when the landing gear is lowered or alternatively only if the pilot selects it.

Chapter 10 deals with linear design issues, and four linear controllers are used to cover the flight envelope. A comparison of multivariable and successive loop closing design is also given to motivate the use of multivariable control. Chapter 11 then deals with discretisation of the control law, and this turns out to be a difficult task as the sampling rate is relatively slow for the bandwidth used. Chapters 12 and 13 deal with pilot command precompensation, and non-linear aspects of the control law design. This builds on the material presented in chapter 3. Chapter 14 deals with non-linear simulation results on the model using TSIM. Chapter 15 then presents piloted simulation and actual flight test results. Finally conclusions are drawn in chapter 16.

CHAPTER 10
LINEAR DESIGN AND ANALYSIS

10.1 Feedback Variables

To a large extent, the two issues of what controls to give the pilot and which measurements to use for feedback stabilisation can be treated separately. Provided that the feedback controller gives sufficient disturbance rejection and damping, flight control mode implementation can be done external to the feedback loop. The approach taken has been to design a multivariable feedback controller using different variables for feedback to those directly commanded by the pilot. This is a different approach to that taken in Part I where, for example, if the pilot wished to control ground speed, then ground speed was used directly for feedback. The three measurements selected for the inner-loop feedback stabilisation are

- Body axis forward acceleration, AXCGB (normalised 'g')
- Body axis vertical acceleration, AZCGB (normalised 'g')
- Pitch rate, QD (degrees/sec).

This choice was made given the following considerations:

- These measurements are generated from accelerometers and gyros in the aircraft itself and have good signal to noise characteristics around the desired bandwidth. Measurements which rely on ground radar (e.g. ground speed) could become unavailable at any stage, and hence cannot be relied on. Aerodynamic measurements are very unreliable at low speeds, and hence cannot be safely used.
- These three measurements can be used across the whole flight envelope, and hence there is no need for any mode changes within the inner loop. This greatly simplifies implementation in that no gain or structure switching is required.
- All three measurements are body axes ones, and hence no problems occur if the aircraft is orientated upside down.

Between VBSTHI and VBSTLO knots a blend to Earth-axes accelerations (AXCGE,AZCGE) is made to reflect that the pilot is interested in controlling the aircraft's speed relative to the ground when in the hover region of the flight envelope.

To each of these three measurements, a proportion of a "hold" measurement is added. For example, a proportion of the pitch attitude measurement is added to the pitch rate measurement to give the new overall feedback variable P :

$$P = QD + \mu_\theta * THETD$$

Feedback using P enables a pitch attitude hold to be effected for steady-state conditions which was a requirement of P6 in section 2.1 . The parameter μ_θ is chosen such that addition of removal of the $\mu_\theta * THETD$ term does not significantly alter the gain and phase characteristics at cross-over. This is illustrated in figure 10.1. The solid and dashed lines represent the open-loop transfer functions from pitch channel error to pitch rate and pitch attitude respectively. The parameter μ_θ is chosen so that g_m is sufficiently large so that the pitch attitude term does not significantly affect the gain and phase properties of $q + \mu_\theta * \theta$ at cross-over. This hold facility could of course be

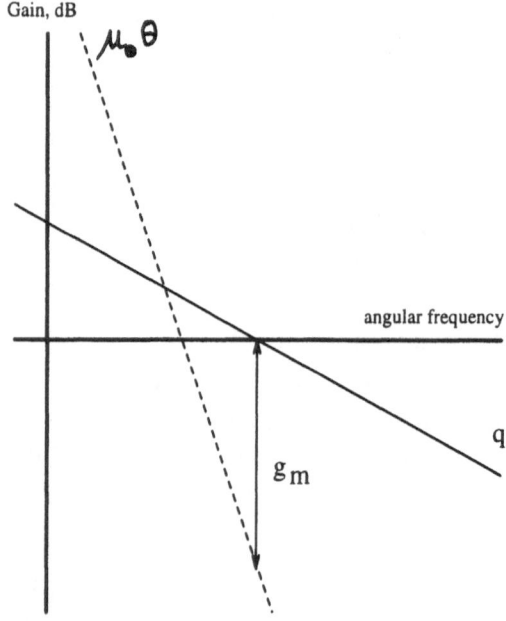

Fig. 10.1. Pitch loop transfer function

effected with a separate outer feedback loop using just the measurement THETD. However, there are advantages to using the augmented feedback variable approach:

− The structure is very simple. No extra compensator dynamics are required.
− Blending of the hold variable to some other measurement is very easy. An example would be to blend out the pitch measurement and replace it with

the flight path signal to give a flight path hold in fully wing-borne flight. This would be effected with:

$$P = QD + \mu_\theta * (\lambda * THETD + (1 - \lambda) * GAMMAD$$

where λ varies between 0 and 1 according to the flight condition.
— The stability of the system is not jeopardised should the pitch attitude measurement become not available. This is more important for the translational feedback variables where the hold variable might come from ground-based radar, and as such is not a reliable signal.

Table 10.1 defines all the feedback measurements across the flight envelope. The Fortran variable XKT2NG converts knots to units of [normalised 'g' × seconds], and FPS2NG converts feet/second to units of [normalised 'g' × seconds]. A term is added to AXCGB to remove the resulting centripetal acceleration picked up by the x-axis accelerometer during a banked turn. This ensures that the aircraft does not slow down in a banked turn.

Table 10.1 Feedback variable specification			
Axis	Stabilisation variable	Hold variable	Ratio of hold to stabilisation variable (λ)
Pitch	Pitch rate, q. Fortran variable QD.	≤ 190 knots: Pitch attitude, θ. Fortran variable THETD. 190–200 knots: Blend between θ and γ ≥ 200 knots: Flight path, γ. Fortran variable GAMMAD	$\lambda_\theta = 1.0$
Horizontal	≤ 105 knots: Earth axes horizontal acceleration in normalised 'g' units. Fortran variable AXCGE 105–125 knots: blend between AXCGE and AXCGB ≥ 125 knots: Body axis horizontal acceleration in normalised 'g' units. Fortran variable AXCGB.	≤ 60 knots: horizontally resolved thrust $\times G_{xv}(s) \times$ XKT2NG. $G_{xv}(s)$ approximates the transfer function from resolved thrust to airspeed. Fortran variable VKTAXF*XKT2NG. Ground speed, VHORKT, if INS signals available 60–100 knots: blend between VKTAXF*XKT2NG and VKTIAS*XKT2NG. ≥ 100 knots: VKTIAS*XKT2NG where VKTIAS is indicated airspeed in knots.	$\lambda_x = 0.125$
Vertical*	≤ 105 knots: Earth axes vertical acceleration in normalised 'g' units. Fortran variable AZCGE. 105–125 knots: blend between AZCGE and AZCGB. ≥ 125 knots: Body axes vertical acceleration in normalised 'g' units. Fortran variable AZCGB.	FPS2NG*$\dot{h}$ where $\dot{h}$ is vertical speed in feet/sec derived from the radar altitude. Fortran variable for $\dot{h}$ is HDOT.	≤ 105 knots: $\lambda_z = 0.25$ 105–125 knots: blend λ_z between 0.25 and 0.125. ≥ 125 knots: $\lambda_z = 0.125$

* The final version of 005 uses a vertical thrust schedule above 125 knots in place of this feedback loop.

10.2 Hover design

In this section the design for one fixed operating point is given, namely for an airspeed of 105 feet/sec (32 m/s) and incidence of 8 degrees. This design is used for airspeeds right down to the hover. Aerodynamics do not have a major effect on the model over this speed range, and so scheduling right down to the hover is not necessary. Note that airspeed measurement is very difficult down at these speeds thus making scheduling difficult anyway. There is a second advantage to performing designs above 100 feet/sec, and this is that the front reaction jet system is not active and hence the reaction control run states do not have to be included in the linear model. This reduces the number of states in the final controller. The linear designs carried out at other airspeeds on the operating envelope use exactly the same procedure.

To reduce the dependency of linearisations for a given airspeed on nozzle angle, instead of working with inputs of thrust magnitude and thrust direction, resolved thrust demands are used i.e. if APF and $NOZZA$ represent thrust demand and nozzle angle demand, then the resolved thrust demands are

$$AXF = APF \times \cos(NOZZA)$$

$$AZF = APF \times \sin(NOZZA)$$

We now follow the procedure of [4] for the design.

1. Scale the inputs. The throttle demand limits are

    $$0.26 \leq APF \leq 1.0$$

    Hence both AXF and AZF vary roughly in the range 0 to 1. The tailplane demand ($ETADA$) can vary in the range

    $$-11.75 \leq ETADA \leq 12.75$$

    and is scaled ×10 so that the demand is now expected to vary roughly in the range -1 to 1. Note that this scaling is only necessary when interpreting closed-loop transfer functions at the plant input - the scaling is lost in the weight W_1 in stage 6.

2. Scale the outputs. The outputs are scaled so that 1 unit of coupling into any of the scaled outputs is as equally undesirable. The horizontal and vertical acceleration outputs are scaled in units of 'g', the acceleration due to gravity. The pitch rate, QD, is scaled ×0.1. The interpretation of this is that a coupling of 0.1g (10 % of the maximum achievable in any direction due to powered lift) is as equally undesirable as the pitch attitude wandering off at one degree per second.

3. Bandwidth requirements and restrictions. The maximum frequency up to which each actuator can be used may be determined by a number of factors including the speed of the dynamics, rate limits and modelling uncertainty. From such considerations the following limitations were arrived at for the hover:

   – Throttle/engine : up to 2 rad/s, the limiting factor being the severe non-linear nature of the fuel-to-thrust characteristic, and the rate limit when spooling up from low engine power.
   – Nozzle : up to 2 rad/s, the limiting factor being the nozzle air-motor backlash.
   – Tail/reaction jets : up to 5 rad/s. The limiting factor here is the high phase roll-off rate above this frequency due to the combined effect of the computational delay, the anti-aliasing filters, the actuation dynamics and the sensor dynamics. Attempting to put in significant phase advance above 5 rad/s would lead to a very unrobust control system in that the combined modelling uncertainty from all these effects would be very large. There is also a lightly damped mechanical control run which back-drives the front reaction jet from the tailplane setting. This also makes it difficult to use bandwidths much above 5 to 10 rad/s when in the hover.

   Note that if different bandwidths are used for the throttle and nozzle, then the bandwidths at which the resolved thrust demands are used will be dependent on nozzle angle. This would thus necessitate making the controller a function of nozzle angle as well as airspeed.

4. Sensor models and second order Padé approximations to represent computational delays and anti-aliasing filters are added to the linear model. For the purposes of design, the sensors, Padé approximations and actuators are cascaded together and approximated with equivalent low order actuator models using balanced truncation. The sensor models can be pulled through to the input like this because all three sensors are modelled with the same dynamics. Extra first order sensor filters with poles at -20 rad/s are added to all three outputs. These were also included within the model reduction. The model reduction keeps the order of the controller down without unduly compromising the robustness of the final closed-loop system. Conventionally model reduction is done on the complete weighted design plant. Here we can not do this as the structure of the weighted plant A-matrix must be invariant between design points so that the scheduling procedure can be carried out. In other words we can only model reduce the dynamics which are invariant between design points when using balanced truncation.

5. Plot the plant singular values and all plant input-output pairs to determine which inputs affect which outputs and whether there are any output directions which are hard to control. Next each of the loops is shaped in turn to have the required low frequency gain, high frequency roll-off, and

a roll-off rate in the region of 20 to 30 dB/decade at cross over. The weight

$$W_P = \begin{bmatrix} \frac{3(s+0.5)}{s^2} & 0 & 0 \\ 0 & \frac{3(s+0.5)}{s^2} & 0 \\ 0 & 0 & \frac{7.8(s+1.5)^3}{s^2(s+10)} \end{bmatrix}$$

is used to achieve this. Figures 10.2–10.4 show the three loop shapes at this point in the design.

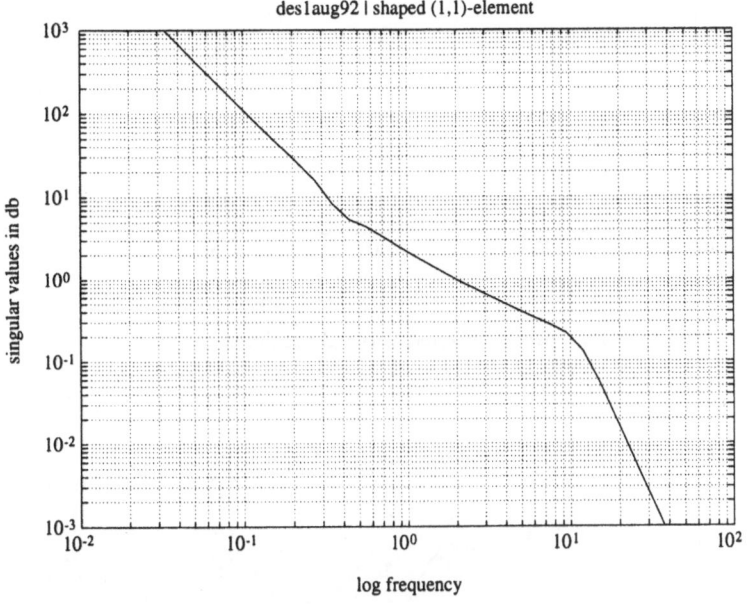

Fig. 10.2. 105 feet/sec design : shaped horizontal motion loop

6. Aligning to achieve desired bandwidths. First a 2 × 2 constant align matrix is found to align the first two singular values at 2 rad/s. Note that for a design in the hover this matrix is just the 2 × 2 identity matrix multiplied by a scalar. As airspeed increases, a step change in horizontal thrust results in an ever increasing vertical acceleration as well as a forward acceleration due to increased lift at increased speed. The align matrix counters this, and hence improves decoupling of the final closed-loop. The align matrix is filled out with a (3,3) entry to give the desired cross-over for the pitch output. If this align matrix is denoted W_A, then the overall precompensator is given by $W_1 = W_P W_A$. Figure 10.5 shows the singular values of the shaped plant with the full order actuator, sensor and computational delay models superimposed on the shaped approximated plant. It can be seen that the approximated plant is close to the

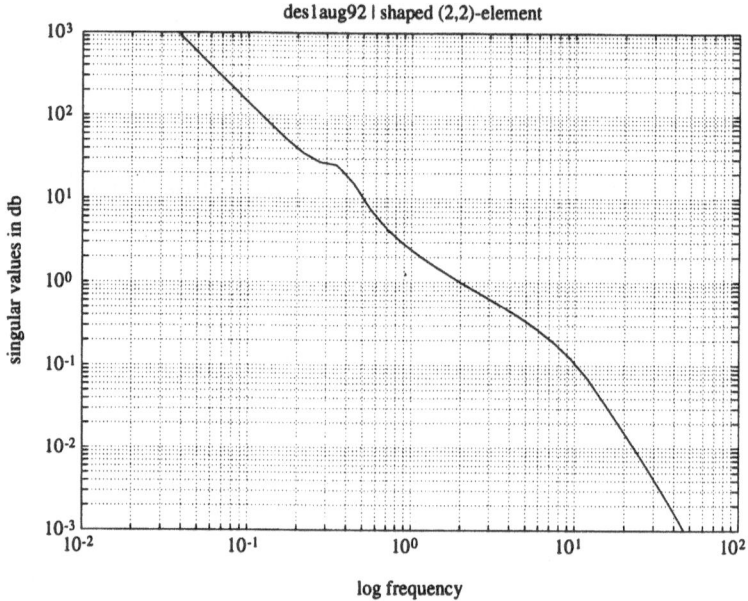

Fig. 10.3. 105 feet/sec design : shaped vertical motion loop

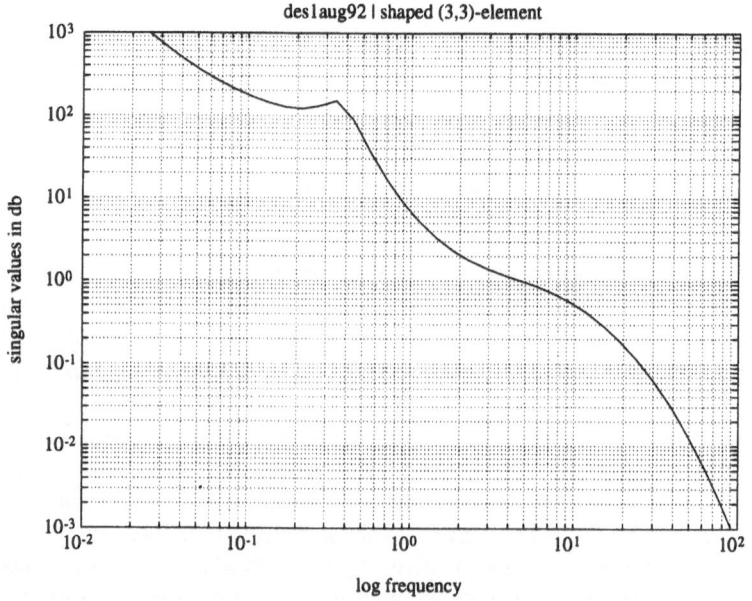

Fig. 10.4. 105 feet/sec design : shaped pitch loop

full order plant around cross-over, and that the specified bandwidths have
been achieved.

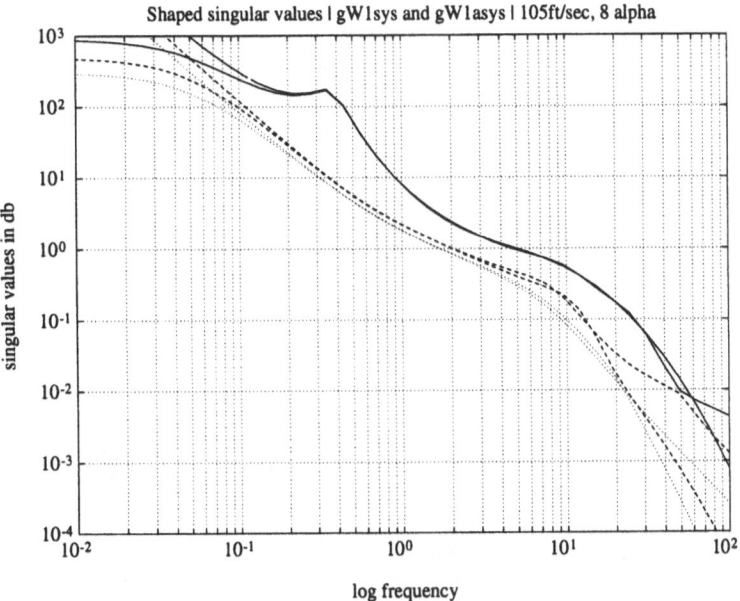

Fig. 10.5. 105 feet/sec design : shaped full order and approximated plants

7. Calculate the optimal controller. The optimal $\gamma_{min} = 1/\epsilon_{max}$ is 2.24 indicating that the specified loop shapes are consistent with robust stability
   ($\gamma < 4$ is usually taken to indicate an acceptable design.) After setting γ
   to 10 % suboptimal, the achieved loop-shaping cost function when evaluated with the full order plant (i.e. not the design plant) is 2.63 as can
   be seen from figure 10.6.
   Figure 10.7 shows the final loop shapes when K_∞ is cascaded with the
   full-order shaped plant.
8. Time response analysis. Figures 10.8–10.10 show responses to step demands on the references. Clearly a high degree of decoupling has been
   achieved, and the time-domain properties look good. The references are
   fed into the loop using the approach proposed in [40]. This is essentially
   implementing the controller as an observer which has an H-infinity filter
   plus LQR static feedback. References are then fed in on the output of the
   state feedback matrix. In [40] is is shown that the closed-loop response
   in this case is equal to $\tilde{N}$ and that the response is both desirable and
   robust to coprime factor plant uncertainty.

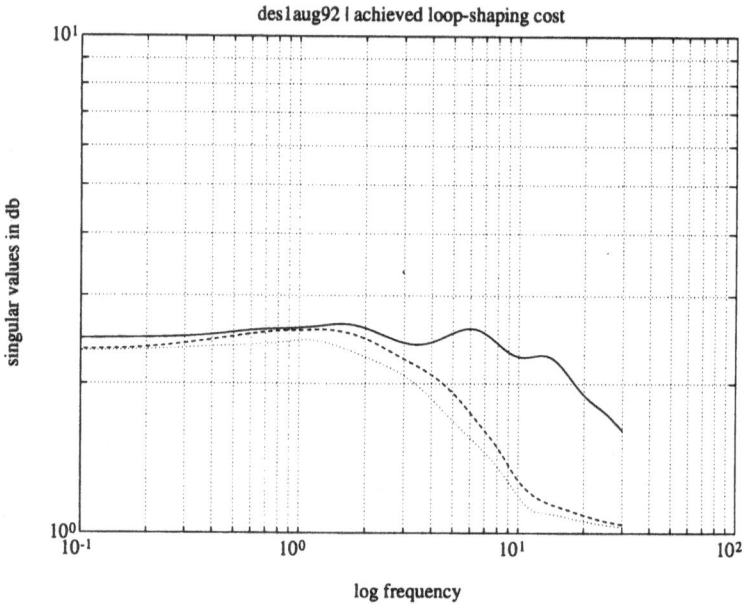

Fig. 10.6. 105 feet/sec design : achieved loop-shaping cost function

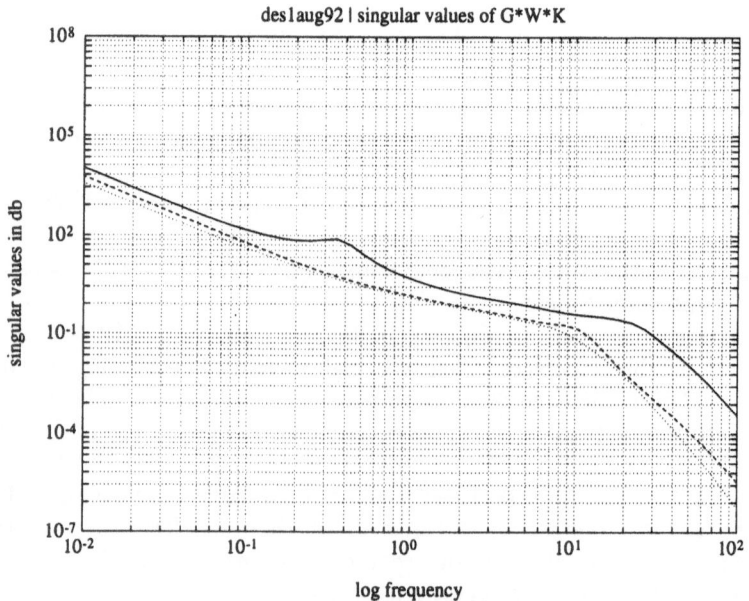

Fig. 10.7. 105 feet/sec design : final achieved loop shapes

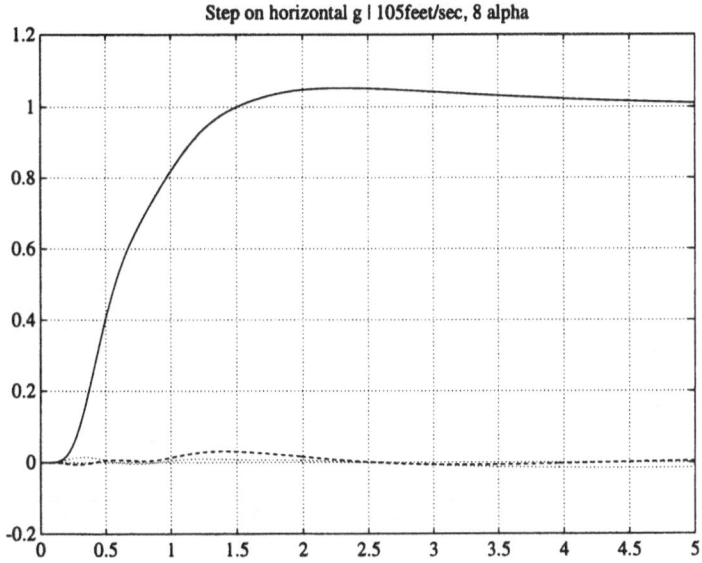

Fig. 10.8. 105 feet/sec design : step demand on forward motion

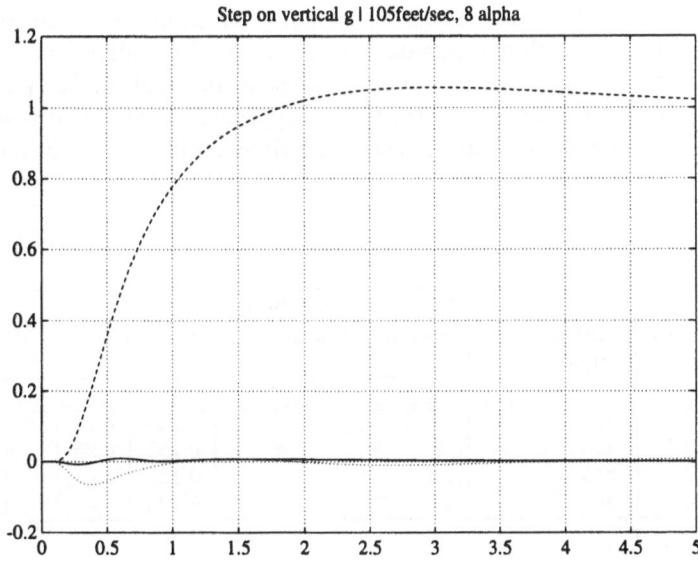

Fig. 10.9. 105 feet/sec design : step demand on vertical motion

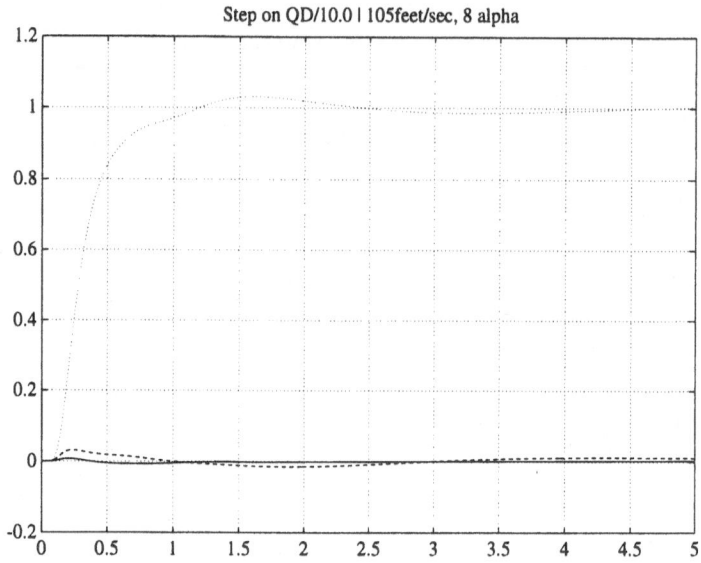

Fig. 10.10. 105 feet/sec design : step demand on pitch motion

10.3 Specifications for the linear designs

Table 10.2 gives the bandwidths used for the three outputs to be controlled for each of the four design points. Also given are the scaling factors for the hold variables. γ is chosen 10% suboptimal so as to avoid the fast pole which appears in the controller as optimality is approached. The model reduction of the design plant prior to the controller design will very likely increase γ too.

Table 10.2: Linear feedback variable specification						
Design	Airspeed (ft/s)	Specified bandwidths			Hold variable multiplier		
		x	z	θ	λ_x	λ_z	λ_θ
1	105	2	2	4.5	0.125	0.25	1.0
2	175	2	2	5	0.125	0.25	1.0
3	245	1.5	0.0	5	0.125	-	1.0
4	345	1.0	0.0	5	0.125	-	1.0

10.4 Justification for using multivariable design

In this section justification for using multivariable design as opposed to successive loop closing is presented. This is an important comparison because the experience of 005 showed that the processor power required for the final control law is approximately twice that required for a loop-by-loop design. This increase in computational requirements therefore needs to be justified in terms of increased robustness or peformance.

The multivariable loop shaping approach is compared with a successive loop closing approach. In this latter approach, each loop is closed using a single-input single-output $\mathcal{H}_\infty$ loop-shaping controller. The bandwidth and weighting function for each loop is kept the same as for the multivariable design i.e. the performance we ask for from each loop is the same as for the multivariable design.

Recall that the $\mathcal{H}_\infty$/loop-shaping robustness optimisation returns a value ϵ which is a measure on the size of uncertainty which can be tolerated. As has already been discussed, this is used as a design indicator, much as gain and phase margin are for classical design. In [40] it is shown that for the single-input single-output case that the following inequalities hold:

$$\text{Gain margin} \geq \frac{1 + \epsilon}{1 - \epsilon}$$

$$\text{Phase margin} \geq 2 \sin^{-1} \epsilon$$

In the multivariable case this design indicator tells us a lot more than the gain and phase margins for the individual loops do in that it allows for simultaneous uncertainty in all loops and for coupling between loops. Figure 10.11 shows graphically the relation between gain and phase margin guarantees and ϵ.

The next sections detail multivariable and successive loop closing designs. The designs are all for one fixed operating point, namely 65 knots airspeed and 8 degrees incidence.

Multivariable longitudinal control

The inputs and outputs are the same as for the design example in section 10.2. Table 10.3 specifies the bandwidths and constants of proportionality for the hold variables.

Table 10.3 : Longitudinal feedback variable specification			
Stabilisation variable	λ	Hold variable	Bandwidth
Horizontal acceleration, g	0.125	Ground speed, $g \times s$	2
Vertical acceleration, g	0.25	Climb rate, $g \times s$	2
Pitch rate, $0.1 \times$ deg/s	0.25	Pitch attitude, $0.1 \times$ deg	5

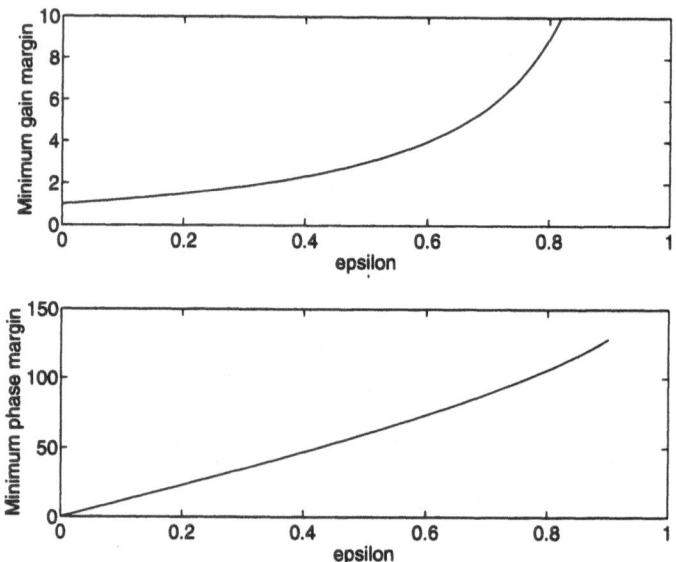

Fig. 10.11. Gain and phase bounds as a function of ϵ

Figure 10.12 shows the shaped open-loop singular values, and figure 10.13 the singular values when cascaded with the $\mathcal{H}_\infty$ controller. Figures 10.14, 10.15 and 10.16 show the time responses. The achieved ϵ was 0.375.

Successive loop closing

The feedback variables, weighting functions and specified loop shapes are identical to those used for the multivariable design. Each loop was closed in turn, and an optimal SISO $\mathcal{H}_\infty$ controller calculated using the loop-shaping / coprime factor stabilisation approach. The pitch loop was closed first, then the vertical motion loop and finally the horizontal motion loop.

Figures 10.17, 10.18 and 10.19 show the time responses. The achieved ϵ's for the three SISO designs were 0.690, 0.667, and 0.543. When translated into guarantees on gain and phase margins using the formulae, the design appears to be very robust (the worst ϵ of 0.543 guarantees a gain margin of 3.4 and a phase margin of 66 degrees). But care must be taken in interpreting these values as they all refer to uncertainty occurring in only one loop at a time. Figure 10.20 shows the loop-shaping cost function (i.e. $1/\epsilon$) achieved when the plant is treated as multivariable and connected to the controller

$$ K = \begin{bmatrix} k_x & & \\ & k_z & \\ & & k_\theta \end{bmatrix} $$

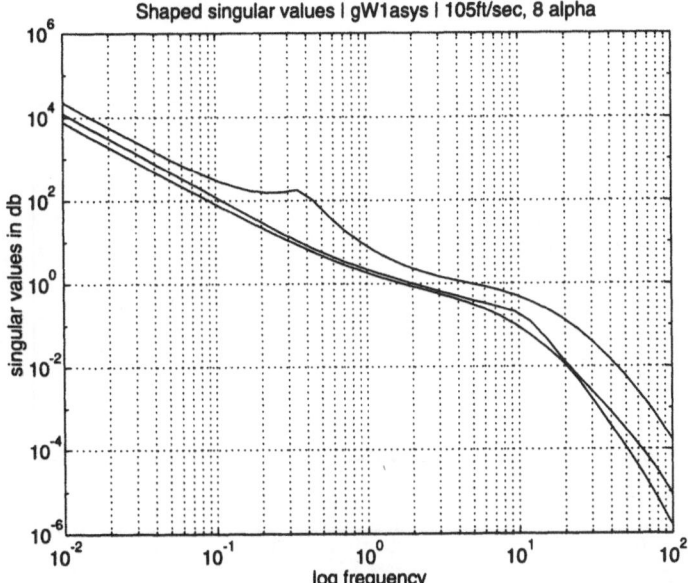

Fig. 10.12. longitudinal multivariable design

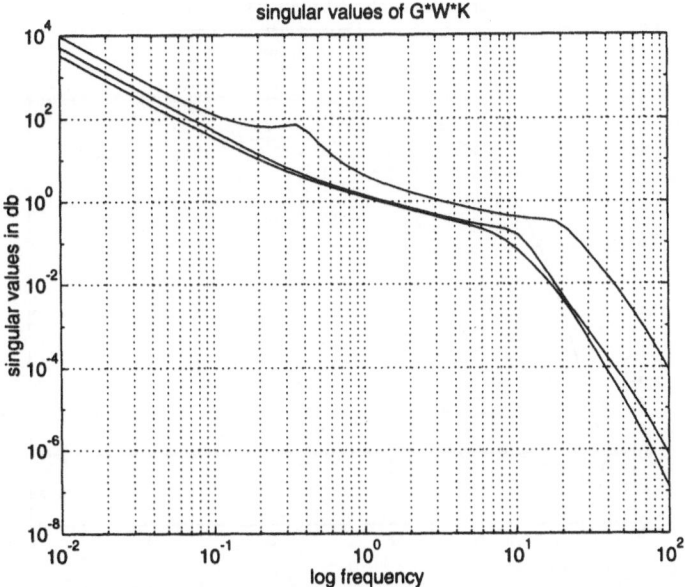

Fig. 10.13. longitudinal multivariable design

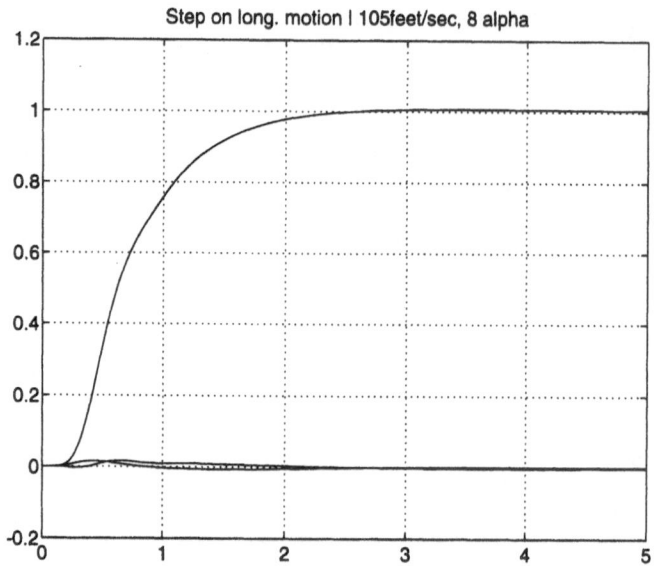

Fig. 10.14. longitudinal multivariable design

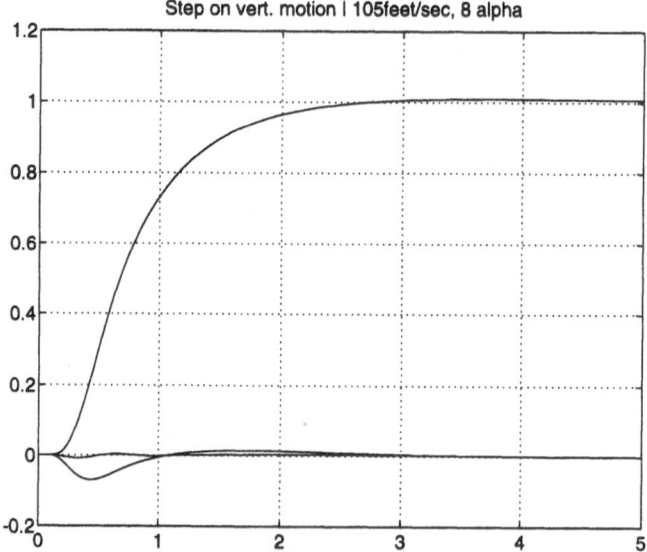

Fig. 10.15. longitudinal multivariable design

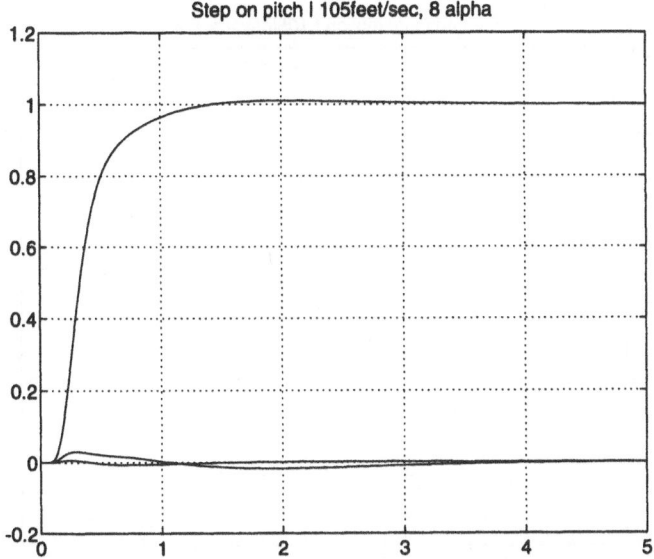

Fig. 10.16. longitudinal multivariable design

where k_x, k_z, and k_θ are the three SISO controller designs. The achieved ϵ for this multivariable analysis is 0.283 indicating significantly less robustness to simultaneous uncertainties in different loops.

Summary of the results

Table 10.4 shows all the stability margins for the various designs. The achieved ϵ's for individual loops of the multivariable designs have been added for comparison purposes.

Table 10.4 : Summary of results				
Design	All longitudinal loops	x-loop	z-loop	pitch
Multivariable	0.375	0.704	0.735	0.546
Successive loops	0.283	0.680	0.667	0.543

As might be expected, the multivariable controller design results in a better achieved multivariable loop-shaping cost function. This is because the design formulation allows for simultaneous uncertainty in all loops. The dif-

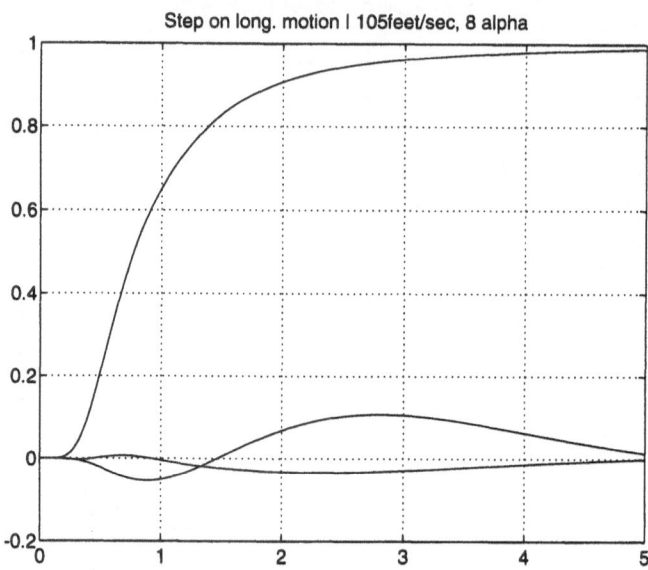

Fig. 10.17. Longitudinal successive loop closing design

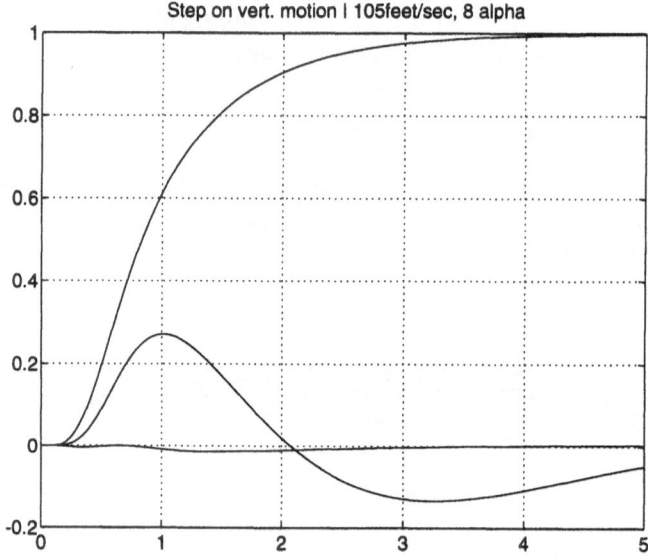

Fig. 10.18. Longitudinal successive loop closing design

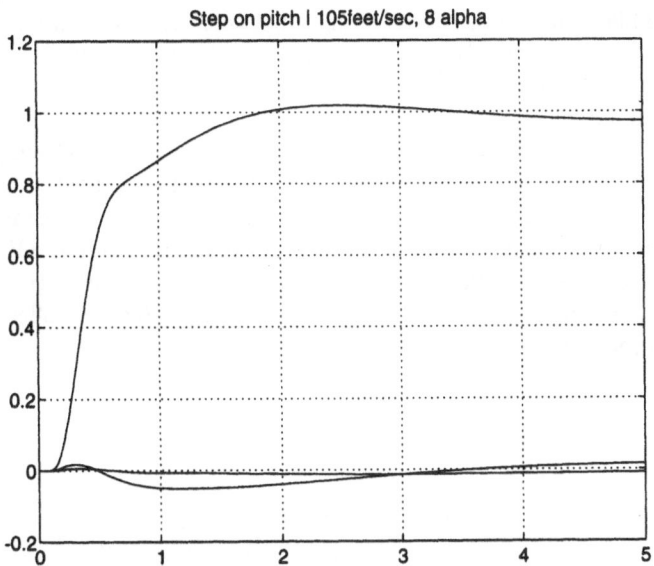

Fig. 10.19. Longitudinal successive loop closing design

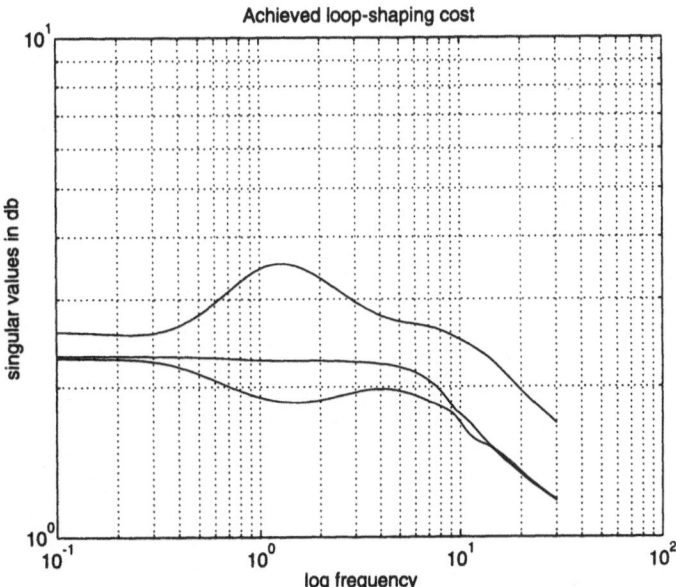

Fig. 10.20. Longitudinal successive loop closing design

ference between the multivariable and successive loop closing designs is substantial. This is due to the high coupling between the horizontal and vertical motion within the plant. Notice also from the table that if uncertainty in only one particular loop at a time is considered, then the multivariable controller also has the edge.

In practice, if one were designing using successive loop closing, the performance of one of the two translational loops would be backed off so as to reduce loop interaction. However, if we are to extract the maximum performance out of the system, a multivariable approach is required. Of course we do not get this extra performance for free in that the complexity of the multivariable controller is higher than the sum of the SISO controllers from the successive loop shaping.

CHAPTER 11
DISCRETISATION

11.1 Sampling at 50 Hz

Figure 11.1 shows the discrete implementation used for piloted simulation. The sampling period was taken to be 20 milliseconds which corresponds to the clock rate of the processors on the DRA XW175 research Harrier. Discretisation was carried out assuming a zero order hold using the Matlab "c2d" command. This method has to be used for the $\mathcal{H}_\infty$ controller as the physical interpretation of the states must remain the same so that the state feedback gains can be linearly interpolated. Tustin based methods do not preserve this structure. Using a zero order hold effectively introduces an extra half period delay into the loop, but simulation showed this to have negligible effect on performance.

Note there are two delays which appear in the block diagram. The first is due to the time it takes to compute the control law, and this is either zero or equal to the frame time of the simulator depending on the order in which the equations are updated. At the start of every frame the control law reads in the measurements and calculates the appropriate actuator demands. The worst case scenario is that these actuator demands are not used to update the aircraft states until the next cycle, and this results in one frame time delay. When the equations are updated this way the simulation corresponds to the situation in the aircraft where the full 20 ms frame time is needed to calculate the controller and thus the delay is unavoidable.

The second delay arises from the Hanus self-conditioned form of the precompensator which relies on having the previous cycle's plant inputs. To arrive at the discretised precompensator, the self-conditioned form of the continuous W_1 was first formed and then this discretised using a zero order hold. The discretisation process does not allow for this second delay when the loop is closed around W_s. However, with the sampling period of 20ms the frequency responses of the continuous and discrete weights were sufficiently close in the frequency range of interest.

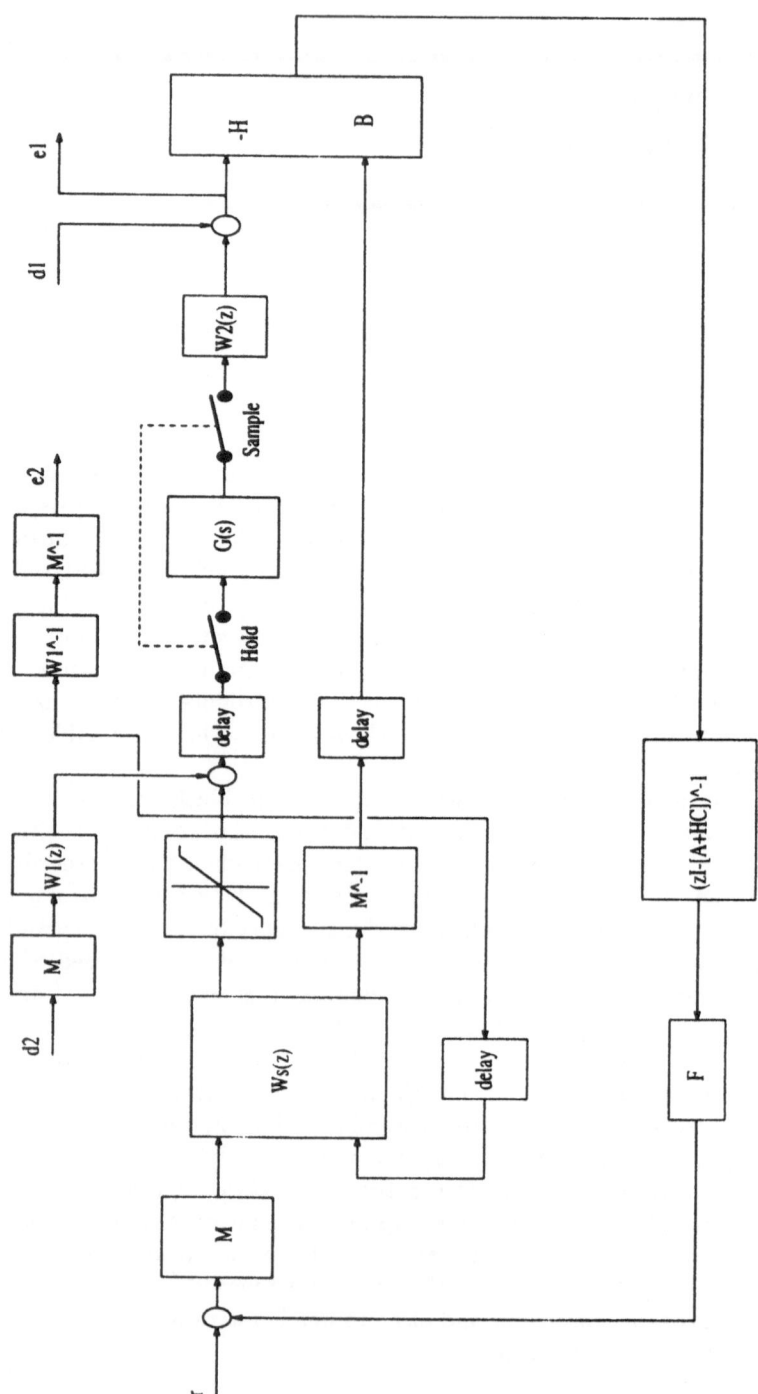

Fig. 11.1. Discrete controller implementation

11.2 Sampling at 25 Hz

The CORAL implementation of the final control law required more processor time than that available on the VAAC processors. The solution to this was to double the sampling period to 40 milliseconds. Designing for this increased sampling period was not just a simple matter of changing the sampling period when carrying out the discretisation. This is because the sampling rate is getting close to the the frequencies of the controller dynamics.

The norm between $[d_1; d_2]$ and $[e_1; e_2]$ in figure 11.1 is the discrete version of loop-shaping cost function used to design the controllers. Figure 11.2 shows this cost function evaluated for the 105 feet/sec design if the controller is discretised using a sample period of 40 milliseconds and zero order holds. There is clearly very poor robustness, and this was substantiated in non-linear simulation where a lightly damped tailplane resonance was observed.

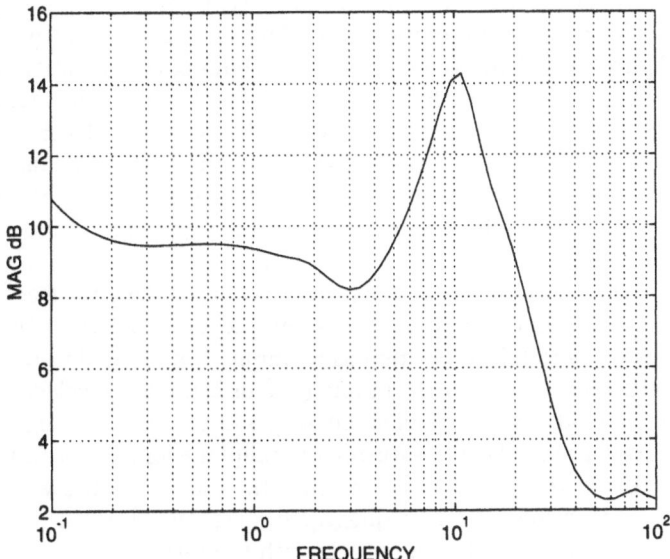

Fig. 11.2. Loop-shaping cost function : 40ms zero order hold discretisation

The pitch loop was identified as causing most of the problem. The cumulative delay due to the zero order hold discretisation of the controller and precompensator (20 ms each) and the computational delays (40ms) means that the phase rolls off very quickly past the bandwidth. The $\mathcal{H}_\infty$ controller counters some of this by putting in phase advance past the bandwidth. As the pitch loop has higher bandwidth than the other two loops, it runs into discretisation problems first.

The solution to finding a robust implementation using the 40 ms sampling required several alterations to the discretisation approach. Firstly, pre-warped Tustin discretisations of the sensor and precompensator were used, the warping frequency being chosen as 5 rad/s which is just past the bandwidth. Sample and hold was then applied to the weighted continuous time plant and direct discrete time design was used to find the observer and state feedback matrices. The discrete time design solution for the loop-shaping is given in [68]. This solution splits into an observer structure as for the continuous time case. The continuous time plant contained second order Padé approximations corresponding to the computational delay. Note that although we could model these delays exactly by appending the discretised plant, this was not done as it would put up the dynamic order of the plant and thus the controller. Modelling the delay with a Padé approximation allowed it to be model reduced along with the actuation models, sensors and weights as discussed in chapter 10.

With these alterations to the way the discrete controller is found we have accounted for all the discrete time effects except for the delay in the self-conditioned form of the precompensator. This delay actually turns out to be critical. Figure 11.3 shows the discrete loop-shaping cost function with this delay, and 11.4 shows it without the delay. The problem essentially arises because the controller poles are given by the eigenvalues of $A + HC + z^{-1}BF$ instead of $A + HC + BF$. This extra delay on the BF term causes a large change to the transfer function of the $\mathcal{H}_\infty$ controller as can be seen from figures 11.5 and 11.6 which show the controller singular values with and without the delay.

The solution taken to this problem was to collapse the pitch feedback loop down so that the observer now has the A-matrix $A + HC + B(:,3)F(3,:)$ and no longer uses the tailplane position as an input. Note that this loses the observer desaturation action for the tailplane. This is not too much of a problem as the tasks already evaluated in simulation and those which might be used in flight testing are unlikely to saturate the tailplane for long periods of time. Note that this is very different to the throttle and nozzle use where operation on the saturation limit is frequent. However, the observer desaturation still operates for the throttle and nozzle with the pitch loop collapsed down. When the control law is off line, the precompensator integrators in the pitch loop are now frozen so as to stop them winding up.

Figure 11.7 shows the discrete loop-shaping cost function with this loop closed for the 105 feet/sec design, and this clearly indicates sufficient robustness. Figure 11.8 shows this cost function across the flight envelope. Remember that below 105 feet/sec the controller is fixed and hence the cost function rises as the hover is approached. A cost function of 4 or less is usually taken to indicate a sufficient amount of robustness, and so fixing the controller below 105 feet/sec looks like a reasonable decision. The discontinuity at 100 feet/sec arises because this linearisation was made around the point at which

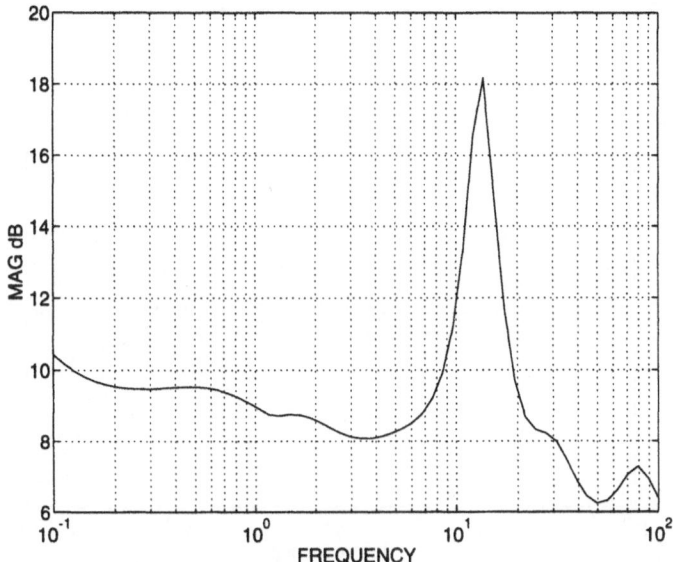

Fig. 11.3. Loop-shaping cost : discrete design

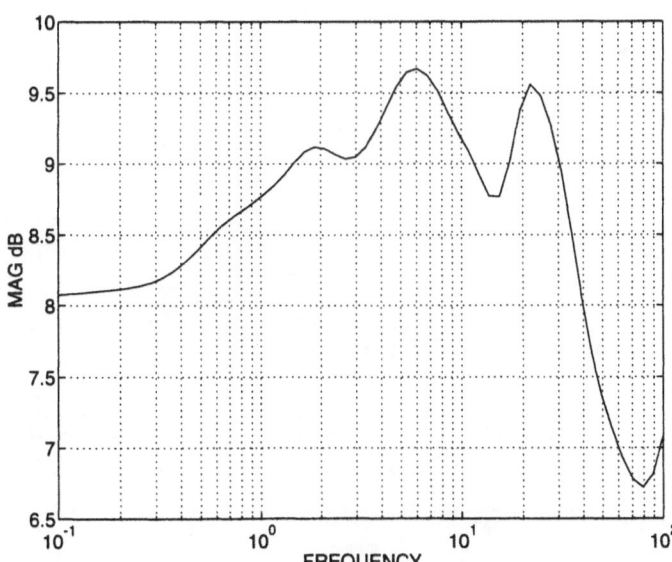

Fig. 11.4. Loop-shaping cost : discrete design, Ws feedback delay removed

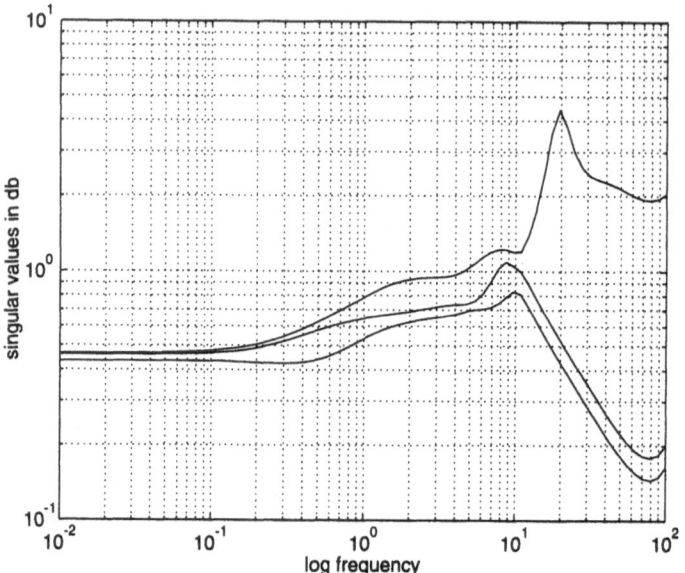

Fig. 11.5. Controller singular values : discrete design

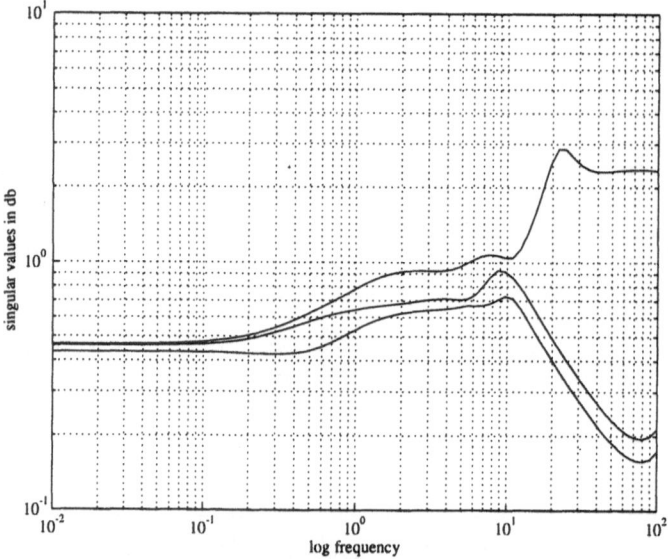

Fig. 11.6. Controller singular values : discrete design, Ws feedback delay removed

the front reaction jet control valve opens. There is therefore a discontinuity in the effect of nozzle angle on pitching moment. Extensive non-linear simulation showed this not to cause a problem when interacting with the control law. The controller is also fixed above 340 feet/sec, and as expected the cost function rises above this speed. The flight envelope for this study is limited to 250 knots (420 feet/sec), and robustness is sufficient in this speed region.

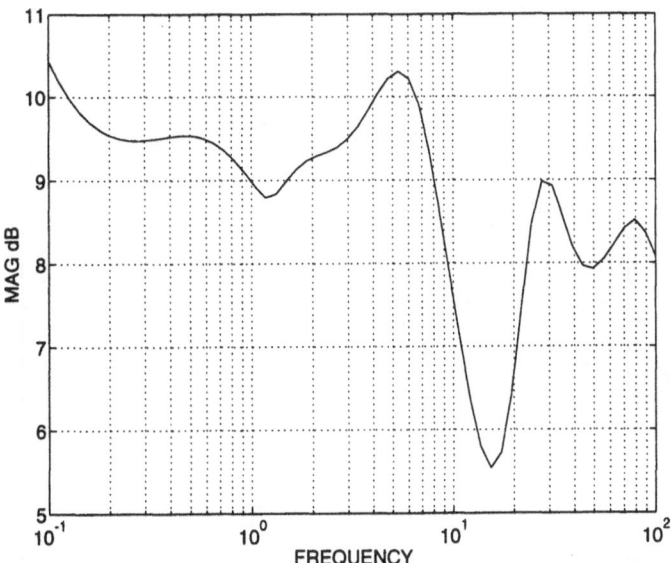

Fig. 11.7. Loop shaping cost : collapsed pitch loop

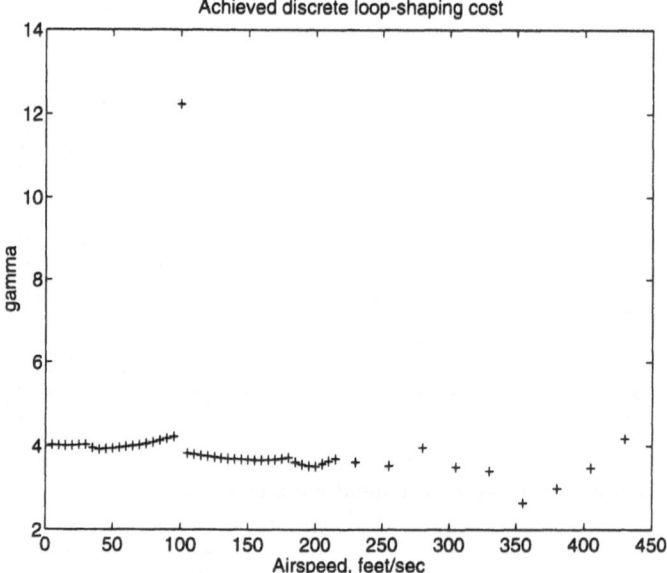

Fig. 11.8. Achieved loop shaping cost as a function of airspeed

CHAPTER 12

FLIGHT CONTROL MODES AND COMMAND PRECOMPENSATION

12.1 Flight Modes For 005

In this section a design specification is drawn up which covers the whole flight envelope. The present control study is restricted to longitudinal control of the aircraft as at present lateral actuation systems are not available to the flight controller on the VAAC research Harrier. However, longitudinal control of the Harrier presents both the greater control challenge and also the greatest potential for reduction in pilot workload. An attempt is made to justify the choices made in the specification alongside its exposition. The specification has been arrived at using a variety of sources including the considerations in §3.3. Discussions with DRA pilots and flight control engineers were highly influential, as were the design study and flight simulations carried out in [4]. Figure 12.1 gives an overall plan of the flight modes.

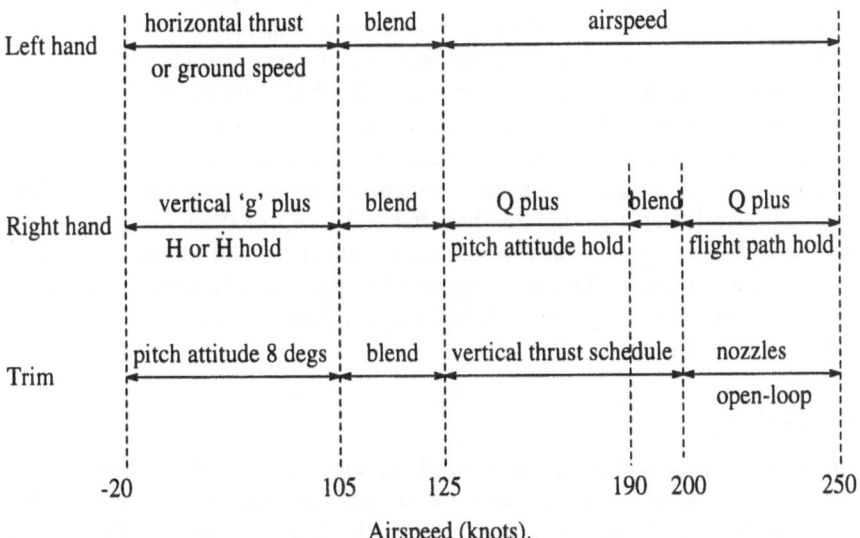

Fig. 12.1. Flight control modes for control law 005

Inceptor 1

For this inceptor the throttle lever in the conventional Harrier cockpit is used, and as such is operated with the left hand. The inceptor is used to command a combination of forward acceleration and forward speed. If $\ddot{x}$ represents forward acceleration, $\dot{x}$ forward velocity, and d the inceptor demand, then

$$\ddot{x} + \alpha\dot{x} = d$$

At high frequencies (short term time response) it looks like an acceleration command, and at low frequencies (long term time response) it looks like a speed demand. In other words, $\ddot{x}$ initially follows a change in d, and then $\ddot{x}$ tends to zero as $\alpha\dot{x} \to d$. This behaves in a similar fashion to the throttle control when in conventional wing-borne flight on the manually flown aircraft; a step change causes an initial acceleration which blends into a constant speed as the aerodynamic drag increases to counter the increased engine thrust. The term $\alpha\dot{x}$ in the demand equation above is thus equivalent to the drag term in the equations of motion for the aircraft. This is consistent with point P1 of the design philosophy of §3.3 in that it gives a conventional response and reflects the aircraft's inherent characteristics.

The possibility of commanding pure acceleration was also considered on the grounds that gradual accelerations and decelerations could be effected without requiring continuous adjustments to the inceptor to maintain the trajectory. This was rejected on the grounds that the resulting aircraft state with hands off the inceptor doesn't tend to a stable equilibrium; even if a speed hold were implemented for centred inceptor (zero acceleration demand), there is still the danger that the detent is missed and the aircraft speed drops off unbeknown to the pilot whose concentration is directed elsewhere.

In the interests of minimising the number of control modes (see P5) this functionality specification is used for the entire flight envelope. Note that the speed-hold for constant control demand makes vertical landing on a carrier easy in that the demand can be set so that the relative speed is zero.

There is one main compromise which has been made with the above choice; in the hover a change in the left-hand inceptor demand gives an unconventional response in that the manually flown aircraft would respond by accelerating vertically. This goes against P4, but the advantage of having a single control mode across the whole flight envelope has been taken to be the higher priority for this design study.

Inceptor 2

The conventional cockpit stick is used for inceptor 2, and is therefore the main right hand inceptor. It has been chosen to command vertical acceleration during hover and low transition speeds. This satisfies **P1** in that the nozzle/throttle combination gives rise to an acceleration over short time periods. This is blended to a more conventional pitch rate demand as soon as there is usable aerodynamic lift.

For speeds between the blend region and 200 knots where powered lift is still required to augment aerodynamic lift, two strategies were tried for setting the powered lift requirement. The first used feedback to control vertical 'g', the vertical 'g' demand being derived from pitch attitude in such a way that the pitch attitude is always the same for a given airspeed when in steady level flight. This scheme has several advantages:

- The scheme is highly predictable - the pitch attitude required to achieve a given climb/descent rate is not dependent on all-up weight or atmospheric conditions as it would be if a vertical thrust schedule dependent on airspeed were used. With the approach here the attitude changes are kept within safe limits.
- Available aerodynamic lift could theoretically be used down to much lower airspeeds than for a scheduled thrust system which would need to make very large attitude changes in the low lift speed regions in order to maintain flight path.
- It essentially keeps angle of incidence within safe bounds for a given range of allowable flight path demands.

The second approach tried was to use a vertical thrust schedule with airspeed. Although this is a less elegant solution, it is much simpler and in practice gave better performance in pilot simulation trials. The first method proved unrobust, particularly at low nozzle angles. Reasons for this are discussed in §13.4.

The mode change is unavoidable given the very different operational requirements and performance possibilities for the hover and conventional flight; in the hover the pilot wants the pitch attitude held constant and to directly command vertical motion, whereas in conventional flight tight control of pitch rate is needed for rapid manoeuvres and pitch pointing. However, the directional functionality (see **P4**) of the controls is always the same in that pulling back on the stick effects a climb in both modes.

For centered stick a flight-path hold is implemented for speeds greater than 200 knots i.e. when the nozzles are not in use. Below the blend region, a vertical speed hold is effected for centered stick. Between the blend region and 200 knots an attitude hold is effected. Ideally this would be a flight path hold, but this can result in the need for large attitude changes at the lower speeds to maintain flight path.

Trim-switch

The trim-switch located on the stick is used to command pitch attitude. As has already been discussed, for hover and transition the pilot would normally like to keep the pitch attitude fixed at 8 degrees i.e. the landing attitude. However, so as not to reduce the operational envelope (**P2**) and to allow the pilot flexibility in achieving a given speed and descent rate (**P3**), the pilot is given the possibiliy of altering this nominal 8 degrees attitude using a trim switch located on the stick. Pushing the trim switch forward effects nose down at 2 degrees per second, and pulling back effects nose up at 2 degrees per second. As the nozzles go aft obviously one of the inceptors must lose authority as only two actuators remain. The trim switch authority is therefore washed out, and pitch rate demand blended across to inceptor 2 (the stick) as discussed above. When the pilot returns to the transition and hover regime, the pitch demand is always reset to 8 degrees.

Height Hold

A height hold facility is also provided. This is engaged by pressing a button on the stick, and dis-engaged by moving the stick away from its central detent i.e. by demanding a non-zero height rate. It can only be engaged when the aircraft is under three-loop control i.e. below the 125–105 knot blend region.

Head-Up Display

The standard VSTOL and Navigational (NAV) mode Head-Up Display (HUD) configurations are used. The NAV-display is centered around a flight path symbol and thus enables the pilot to set the velocity trajectory to where he wishes the aircraft to go. The VSTOL-display is centered around a symbol indicating the pitch attitude of the aircraft to the pilot so that the landing attitude can easily be set. Superimposed on this is a flight path symbol which is of course of diminishing usefulness as the aircraft approaches the hover. Originally the change in HUD mode was used to signal to the pilot the point at which the stick blend from pitch rate to vertical 'g' starts during a deceleration. However, the VSTOL display was not liked up at these airspeeds as the pitch attitude tends to dominate the display whereas the pilot is still controlling flight path. The display change was thus lowered to 60 knots. The pitch attitude symbol was also removed from the VSTOL display, and the flight path / height rate symbol increased in size so that it matches the NAV display symbol. Pitch attitude can be displayed in the hover using the incidence marker along the left-hand edge of the display.
The speed signal given in the circular window at the top left of both displays shows the following:

– Horizontal component of measured (indicated) airspeed in the xz-plane whilst the stick is commanding vertical 'g'. If the INS ground speed signal is available, then this is shown instead.
– Indicated airspeed whilst the stick is commanding pitch rate.

Thus if there is no ground speed signal available, the displayed speed is an approximation to ground speed when in vertical 'g' mode and in the absence of wind. The speed demand index is calibrated such that 12 o'clock represents 0 knots, and 6 o'clock represents 250 knots demanded airspeed. Each of the graduations around the dial thus represents 50 knots. This set-up ensures that there is no confusion as to demanded speed which there would be if 6 o'clock were chosen to represent 100 knots. It also means that the inner revolving indicated airspeed hand (which is calibrated at 100 knots for each half turn) lines up with the demanded speed symbol at 250 knots which is the airspeed at which it is envisaged the control law would be brought on-line. All other symbols on the HUD have their usual interpretation. For hover operation a finer speed demand scale would be beneficial.

12.2 The left-hand inceptor

Figure 12.2 illustrates the precompensation of the throttle lever demand XTHRTS to generate the closed-loop demand DEMX. Non-linear scaling is used so that -90 through 0 degrees corresponds to -10 through 100 knots airspeed, and 0 through 90 degrees corresponds to 100 through 500 knots. This non-linear characteristic allows finer airspeed adjustments at low speed.

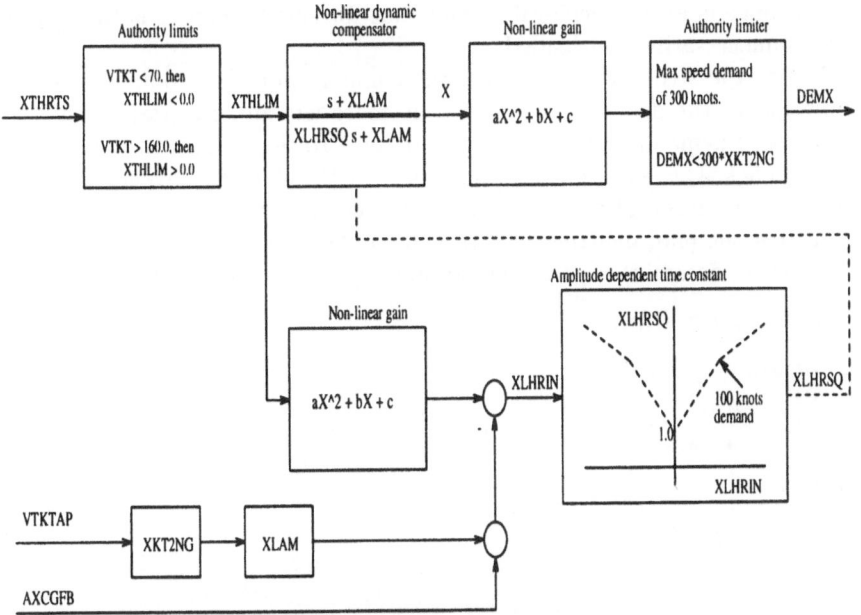

Fig. 12.2. Left-hand inceptor demand precompensation

In addition to the non-linear scaling, the demand is also put through a non-linear dynamic compensator. This dynamic compensator is essentially an amplitude dependent lag; if the difference between the unlagged demand and the current value of the feedback variable is small (i.e. XLHRIN is small), then the lag is small. Conversely, for large demands the lag is large. The reason for having this characteristic is that without the lag, a large change in speed demand would cause large abrupt changes in the engine rpm. Not only is this bad for the engine, it also gives an unconventional response (where lever setting essentially sets engine rpm). For smaller changes in speed demand, the pilot will want a rapid response, particularly during low speed operations. Hence no or little extra lag is required for small demand changes. Note that for small demand changes, the corresponding change in engine rpm is small, and so rapid rpm changes are less of a problem.

Two variables are used to set the degree of command damping. XLHRS1 sets the rate of increase of lag with demand size for demand changes up to 100 knots, and XLHRS2 for demand changes larger than 100 knots (refer to the amplitude dependence block in figure 12.2). These values were set during piloted simulation to 1.0 and 0.4 respectively. First XLHRS1 was set so as to give a suitable response (i.e. smooth engine response, but sufficiently fast) when moving the demand abruptly from the hover stop to the centre detent i.e. 100 knots airspeed demand. Then XLHRS2 was set so that putting the throttle lever fully forwards from any starting speed within the flight envelope (i.e. < 250 knots) just causes the engine to saturate i.e. it is ensured that full performance is always achievable, and that for smaller demands the engine is not unnecessarily over-loaded.

Notice the structure of the dynamic compensator. The compensator actually makes the demand XTHRTS look more like an acceleration and less like a speed demand as XLHRSQ is increased. To see this, recall that we are feeding back $\lambda_x \dot{x} + \ddot{x}$ which, as discussed in §2, gives a conventional type response i.e. an initial acceleration which blends into a long-term speed demand. Let the pilot demand be $r(s)$, and let the demand precompensator be $D(s)$. Then, assuming that the closed-loop dynamics are very fast,

$$s^2 x + \lambda s x \simeq D(s) r(s)$$

and hence

$$x \simeq \frac{D(s)}{s^2 + s\lambda} r(s)$$

. If we put

$$D(s) = \frac{s + \lambda}{\epsilon s + \lambda}$$

then

$$x \simeq \frac{1}{s(\epsilon s + \lambda)} r(s)$$

which corresponds to feeding back $\lambda_x \dot{x} + \epsilon \ddot{x}$. Hence ϵ (i.e. XLHRSQ) effectively alters the acceleration component of the demand.

12.3 The right-hand inceptor

Figure 12.3 shows the precompensation of the stick demand (i.e. the right-hand inceptor) prior to the closed-loop.

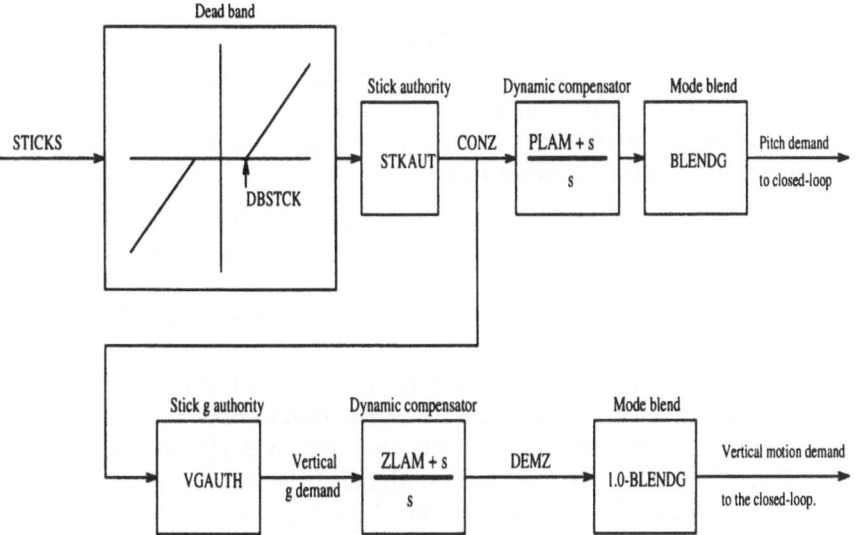

Fig. 12.3. Precompensation of the right-hand inceptor

The first block is a dead-band which allows for the slack in the stick about its central position. When the stick is in this dead-band, the either the flight path hold or the HDOT-hold is active (depending on airspeed). The variable STKAUT (literally "stick authority") is used to set the sensitivity of the pitch rate response for a given stick deflection. VGAUTH affects only the vertical 'g' demand, and is set to give the desired stick sensitivity when in the hover once STKAUT has been set for wingborne flight operation. The dynamic compensator is required to convert the pilot vertical 'g' demands (i.e. $\ddot{z}$) into closed-loop demands where $\ddot{z} + \lambda\dot{z}$ is the feedback variable. Let this dynamic compensator be denoted by $R(s)$, and the pilot demand by $r(s)$. Then

$$s^2 z + \lambda_z s z = R(s)r(s)$$

Putting

$$R(s) = \frac{s + \lambda_z}{s}$$

gives

$$s^2 z = r(s)$$

as required. The integrator is frozen if any of the following conditions occur:

- Vertical thrust authority runs out. For example, if the pilot pulls back in the stick in the hover and holds it there, eventually the engine will saturate high. When this happens, there will be a mismatch between demanded and achieved vertical acceleration, and hence the demand integrator must be frozen.
- The throttle hits its lower limit when in the powered lift regime. Freezing the integrator in this case stops the command integrator winding up when the pilot demands a descent rate which cannot be achieved.
- The descent rate limiter cuts in. Again this stops a mismatch between stick demand and achieved descent rate.

The variable BLENDG is used to blend the stick from demanding pitch rate in at high speed and vertical 'g' in the lower end of the powered lift regime. The region of this blend is set by the two speeds VBSTHI and VB-STLO which following piloted simulation trials have been set to 125 and 105 knots respectively.

The flight path and vertical speed holds are automatically implemented via the choice of feedback variables used. The change between attitude hold and flight path hold over the 190–200 knot region is effected by blending the hold variable from pitch to flight path using the variable BLENDN. The command path dynamics shown in figure 12.3 do not change.

12.4 Trim-switch and pitch hold

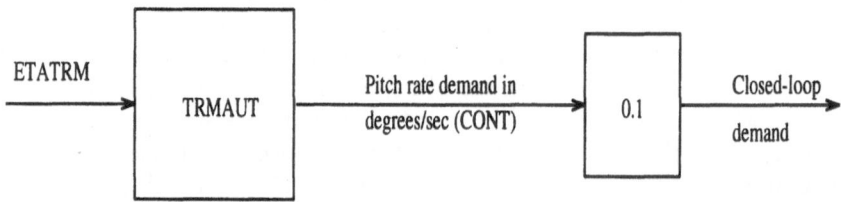

Fig. 12.4. Trim-switch command conditioning

Figure 12.4 shows the conditioning of the trim switch signal. TRMAUT is nominally set to 1.0, and this results in the aircraft pitch attitude varying at a rate of +/-1° per second when the trim switch is pressed. The scaling factor of 0.1 is required because the pitch feedback variable is 0.1(QD + PLAM*(THETD-DEMPT)) where DEMPT is the pitch hold demand; when the trim switch is pressed, DEMPT is set equal to THETD which effectively means that the feedback variable is 0.1QD, and hence a pure pitch rate is commanded by the trim-switch. When the trim-switch is released, DEMPT is frozen at the current pitch attitude value thus giving the pitch hold function.

CHAPTER 13

LIMITERS AND NON-LINEAR ASPECTS

13.1 Engine compensator

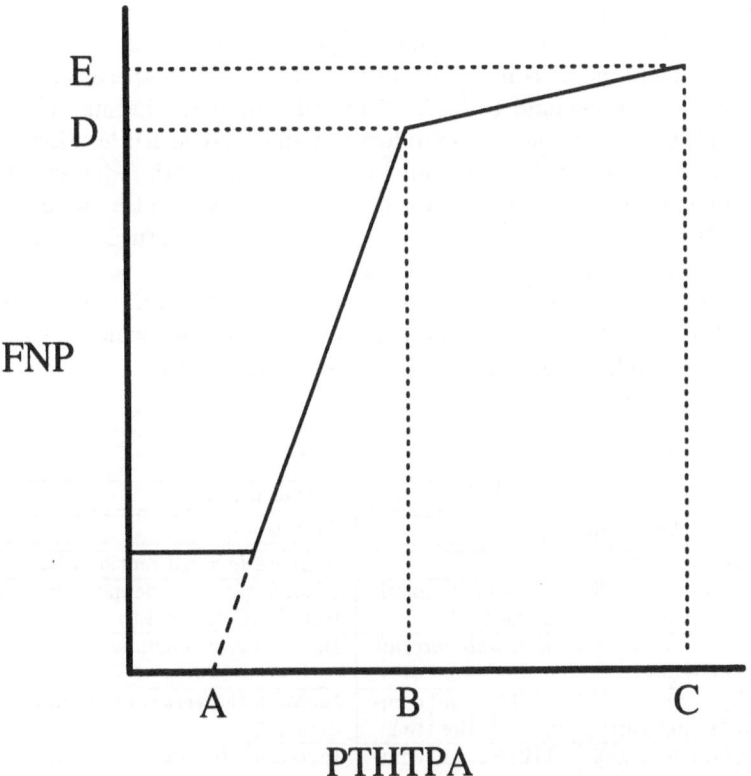

Fig. 13.1. Schematic representation of the engine characteristic

Figure 13.1 shows the relationship between fan speed and throttle servo demand. The severe non-linearity is due to the engine governor cutting in at high fan speeds. Note, that as well as the change in gain there is also a change in the dynamics, they being much faster at high engine speeds. An engine controller has been developed to compensate for this large non-linearity. Refer to the block diagram in appendix B where the engine controller is shown within a dotted box. It gives an approximately linear relationship between the demand APF and achieved fanspeed (which in turn reflects achieved thrust). Note that the compensator relies on knowing the fan speed at which the governor cuts in which in practice will not be known exactly. The engine controller has been left parametric so that the values can be tuned to the actual aircraft if necessary. FORTRAN variables ENJA, ENJB, ENJC, ENJD and ENJE correspond to the values A, B, C, D and E in figure 13.1.

13.2 Desaturation scheme

Control law 005 uses the observer desaturation scheme supplemented by the prioritized desaturation scheme during specific flight conditions. The use of the prioritized scheme is necessary because on occasions there is a choice as to which control demand to back off in order to stop wind-up. The classic example is if the pilot demands full forward and vertical acceleration when in the hover: there is insufficient engine thrust to meet both requirements, and the available thrust must be apportioned sensibly between the two directions. Table 13.1 gives the specifications for the high-gain desaturation implementation. Note that so as to have priority over the observer scheme it cuts in at 95% of full actuator authority by using artificial software saturation limits. PTHTPL is the software generated throttle low limit which ensures that engine rpm are kept to a safe level as airspeed decreases.

Table 13.1 - Desaturation priorities	
Saturation condition	Action
Nozzle $> 95.5°$	Increase forward motion demand
Throttle $> 95\%$ max and commanding vertical 'g' with the stick	Disable the stick demand integrator (set ENABLZ to 1.0)
Throttle $> 95\%$ max and aircraft banked over	Back off z-axis demand
Throttle $< 1.05*$PTHTPL and commanding vertical 'g' with the stick	Increase the vertical motion (climb) demand
Throttle $< 1.05*$PTHTPL and commanding 'q' with the stick and the nozzles have an aft horizontal component	Increase the forward acceleration demand

In addition to the above strategies, logic is also placed external to the loop to limit pilot demands in certain situations. For example, below 100 knots, the authority of a forward speed demand is limited so as to limit the engine power diverted from controlling the vertical motion.

Models of the actuator rate limits are built into the controller, and thus the observer scheme effects a desaturation action on them. Refer to the overall block diagram in appendix B to see how this is done.

13.3 Limiters

13.3.1 Descent rate limiter

A descent rate limiter becomes active as the stick demand is blended from pitch rate to vertical 'g' demand mode. This is necessary because of the very small power margin when in the hover regime which means that large descent rates take a long time and a lot of work from the engine to counter. The value of the descent rate limit is set to 1000 feet/minute. Note that this system does not give totally care-free handling - it only prevents the pilot demanding excessive descent rates when in the vertical 'g' demand mode. It is still possible to build up large descent rates during deceleration from wing-borne flight by holding the nose down too far at speeds where aerodynamic lift runs out. Note that this also true of the manually flown aircraft and so should not present a problem.

13.3.2 Low throttle limit

A lower limit on the throttle cuts in when in the powered lift regime. This is necessitated by the considerable lag in the engine response when at low engine revs. Note that this limit does not compromise the existing flight envelope in that the pilot would never throttle fully back when in the powered lift regime.

The airspeed region in which the limiter is blended in is between 200 and 114 knots. The lower limit on the fuel flow demand is set to 0.67 which is just above the low fuel flow safety limit in the VAAC Harrier Throttle-Nozzle Monitor (TNM).

13.3.3 Inverted flying protection

If the aircraft is rolled or executes a loop, then feedback of signals such as pitch attitude or flight path which are measured in Earth axes cannot be used for feedback. Hence the flight path hold facility is blended out between 70 and 80 degrees pitch attitude or bank angle. This is very simply effected by multiplying the hold term in the pitch feedback loop by the variable (1.0-AINVER) where AINVER normally takes the value 0.0, and blends to 1.0 at high bank and pitch attitude values.

13.4 Thrust/attitude trim algorithm

Refer to figure 12.1 where all the flight control modes are described. Because we are using a two-inceptor demand system we need some strategy to define how to use the third degree of freedom. Two different schemes were tried for 005, one using feedback and one using open-loop thrust demand setting.

The first designs used the feedback approach to provide an "auto-trim" function. The vertical motion demand is back-driven from the current pitch attitude, and is such that if the pitch attitude is ZEROZZ degrees then zero vertical acceleration results. ZEROZZ is a function of airspeed, it being 8 degrees at 100 knots and 2 degrees at 200 knots. Thus the attitude setting for a particular flight path is always the same - it is not a function of aircraft weight or atmospheric conditions as would be the case for a pure thrust schedule with airspeed.

The following line of code generates the back-generated closed-loop vertical motion demand which is denoted TH2DMZ:

TH2DMZ = -XKT2NG *
      VKFILT*ZLAM*sin(DEGTOR*SCLALF*(FILTPD-ZEROZZ))

Essentially the VTFILT*sin(x) generates a vertical motion demand in knots from the flight path demand DEGTOR*SCLALF*(FILTPD-ZEROZZ). The variable DEGTOR simply converts degrees into radians, and SCLALF is an authority variable which can be set to suit the pilot. FILTPD is the filtered pitch signal (pitch attitude and pitch rate combined with a complementary filter with pole at -1 rad/s). With SCLALF set to 1.0, a one degree change in pitch attitude gives a one degree increase in flight path demand. This in turn means that incidence is unchanged, and that the increased flight path is effected by and increase in engine work. Note that the dynamic response is faster than if just the engine were used i.e. keeping the pitch the same, increasing the engine load and thus decreasing alpha. If we make SCLALF larger than one, then less engine power and more aerodynamic lift is used to change the flight path. However, too much reliance on an increase in in aerodynamic lift will result in very large pitch attitude changes at lower airspeeds in order to achieve the desired flight path.

In practice this scheme did not function as well as was hoped. Three reasons are suggested for this:

1. Large pitch attitude changes were required in the deceleration to maintain flight path. This is partly because the vertical thrust demand changes resulting from changes in attitude were relatively slow, and thus large pitch changes are required for a fast flight path response.
2. With the flight path hold engaged the pitch loop uses flight path as the hold variable. The pitch attitude then sets the vertical speed demand to the vertical speed hold loop. In other words we have hold variables of vertical speed and flight path simultaneously, and as one is dependent on the other performance is likely to be poor. This turned out to be true in

practice in that a slow mode resulted in the form of the nozzles taking a long time to settle after a flight path demand change.

3. There is a large pitching moment generated by the nozzles moving forward and the engine rpm building up. This causes pitch transients which in turn give rise to flight path transients. With the experience gained with 005 it is thought that better performance can be achieved by controlling pitch attitude tightly with an inner loop, and then controlling flight path with a lower authority outer loop.

The second approach tried used an open-loop vertical thrust schedule. This leaves flight path to be controlled by pitch attitude changes alone. The vertical thrust schedule is nominally zero at 200 knots and 93 % of max fan speed by 120 knots. A correction term which is a function of the aircraft forward acceleration is also used to ensure that overall lift is the same for a given speed regardless of whether the aircraft is accelerating or decelerating. This approach gave much more predictable results than the solution using feedback. However, the flight path hold had to be replaced by a pitch attitude hold for the 125–190 knot speed region. Again, as with the feedback approach, unacceptably large pitch attitude changes resulted from a flight path hold system over this speed region. A scheduled pitch attitude demand is superimposed upon the pilot's attitude demand to give an approximate flight path hold over this speed region.

13.5 Forward nozzle in wingborne flight

If the throttle lever is set more than three quarters back from the central detent then the nozzles are are brought fully forward, and the engine power increased to maximum. This gives maximum braking power. Three quarters back on the speed demand corresponds to a demand of -10 knots, and so a normal decel to hover is still effected by demanding zero knots.

13.6 Hand-over logic

The variable IFLY is used to signify switching control between front and rear cockpits in the aircraft. The variables PTHTPX, THEJDX, and ETADX are the servo demands generated by the control law. The variables PTHTPA, THEJDA, and ETADA are the actual demands to the servo demands, whether generated from the control law or the safety pilot. The effect of IFLY is as follows:

– IFLY=0 : the model is flown from the desk using the control law.
– IFLY=1 : the model / aircraft is flown from the simulator cockpit / aircraft rear cockpit using the control law.

– IFLY=2 : the model / aircraft is flown manually from the desk / aircraft front cockpit.

In simulation, by setting IFLY = 2 and flying manually from the desk, handover to the control law is simulated by then setting IFLY=1 and getting the pilot in the cockpit to take control.

13.7 Feedforward of bank angle

If the aircraft banks over in the hover, then the vertical thrust must be increased to maintain height. If left to its own devices, the control law would do this using feedback. However, a much quicker dynamic response can be obtained using feedforward from bank angle to engine demand. To achieve this effect the engine controller demand APF is divided by

$$(AMAX1(COS(DEGTOR*PHID*BNKAUT),AMAX1(0.5,BLENDG)))$$

The term BLENDG ensures that this feedforward term is removed once out of the hover in that it is set to 1.0 when commanding pitch rate with the stick. The 0.5 term is to ensure that divide by zero never occurs. BNKAUT sets the authority of the compensation, and is set during piloted simulation such that height is neither lost nor gained during a small banking manoeuvre.

In wing-borne flight a second feedforward term from bank angle to airspeed demand is used. This was found necessary as the speed loop had insufficient authority to hold speed during steep banked turns. The authority of the feedforward is set by BNKFIX, nominally set to 1.5.

13.8 Height-hold

Following piloted simulation a height-hold facility was added. The rationale behind this is discussed in the piloted simulation section. It is very simply effected by adding an extra term HACTIV*(H-HDEM) to vertical acceleration feedback variable where H is the height measurement, HDEM the demanded height which is set when a button on the stick is pressed.

13.9 Feedforward from nozzles to tailplane

Changes in direction and magnitude of the thrust vector cause large pitching moments. During the deceleration to hover the rate of change of this thrust vector is high, and this results in pitching moments which the control law is unable to adequately counter. To reduce the effect the following feedforward term is added to the the tailplane demand:

FFTH2P = FFPAUT*FNP*THEJDX*0.01

FFPAUT is set in simulation so as to minimise the effect on attitude from changes in thrust vector.

13.10 Performance measurement points

Disturbance entry points into the loop have been added for use during simulation. The disturbances enter the loop at the input and output to what is the weighted plant used for design. Thus by looking at the appropriate closed-loop errors generated for given disturbances, the achieved coprime factor cost function can be evaluated. This in turn gives an indication of the achieved robustness on the non-linear model. Referring to the overall block diagram in appendix C, the disturbance inputs are DISTI1(*) and DISTI2(*), and the corresponding closed-loop errors are DISTO1(*) and DISTO2(*).

13.11 Landing logic

The measurement WTONWH (WeighT ON WHeels) is assumed available. For the purposes of piloted simulation, the variable is set to 1.0 when the height above ground is 8 feet or less.

If the aircraft is on the ground, and thus WTONWH = 1.0, then:

- The prioritized desaturation scheme is turned off i.e. DESATU(*)=0.0 . This leaves the observer desaturation scheme still in place.
- The pitch demand, DEMPT, is set to 6 degrees. This is reset to 8 degrees just after take-off.
- The tailplane is set to 0.0 degrees to stop it wandering off to its end stop.
- The low throttle servo limit is reduced to 0.26 so that the engine can fully wind down.

13.12 Implementation in Fortran and Coral

The control law was implemented in Fortran for simulation on the non-linear model and on the DRA flight simulator. As all the linear controller designs were implemented in discrete form, no integration algorithms were needed to implement the control law.

Implementation in Coral required more care. The discrete equations are set for updating every 40 milliseconds. As the software environment calls the control law every 20 milliseconds, logic was placed around the code which ensures that its implementation is divided over two cycles. The conversion of the FORTRAN code into CORAL is documented in [69]. The VAAC Harrier has three 8086 processors with 8087 co-processors available to the control law. The software environment is clocked at 20 milliseconds sample time, and this results in approximately 14 milliseconds being available to the control law on each processor. Table 13.2 summarises approximate computation times for these processors assuming single precision floating point operation for all calculations.

Table 13.2 – Worst case time requirements	
Operation	Processor time
Addition	17 μs
Subtraction	17 μs
Multiplication	19 μs
Division	39 μs
Trigonometric functions	19 μs
Square root	36 μs

Timings of the complete control law on a single processor resulted in a total update time of approximately 80 milliseconds. This figure could be reduced by dividing the matrix interpolation task (i.e. the scheduling) over several cycles. However, there is a limit to how much computation time can be saved this way and it was not possible to fit the control law implementation into the available computation time. The solution was to double the sampling period for the control law, and to divide the control law update over two 20 millisecond frames. Having done this, the control law was divided as follows:

- Processor 1
  - Frame 1. Calculate $x_1 = Ax_n$ where A is the 20 A-matrix and x_n the state vector.
  - Frame 2. Calculate $x_2 = Bu_n$ where u_n is the observer input. Calculate $x_n = x_1 + x_2$. Calculate $y_n = Fx_n$.
- Processor 2
  - Frame 1. Interpolate the first half of the 20 × 20 A-matrix.
  - Frame 2. Interpolate the second half of the 20 × 20 A-matrix.
  - Frame 3. Interpolate the 20 × 6 B-matrix, 3 × 20 F-matrix, 3 × 3 M-matrix (align matrix) and the 3 × 3 D_c-matrix (low frequency closed-loop gain matrix).
- Processor 3
  - Frame 1. Filter and process all measurements. Process all pilot demands. Precompensator calculations. Calculate the observer input, CTRLU.

– <u>Frame 2.</u> Calculate and limit all actuator demands. Send signals for data logging via the telemetry link.

Communication via global memory (i.e memory available to all three processors) can slow down the overall implementation significantly. However, processors 1 and 2 have shared memory, and hence with the above arrangement the interpolated state-space matrices are simultaneously available to both the state equation updater and to the scheduler. Communications between processor 3 and the other two are minimal, and consist of passing CTRLU(6) and CTRLO(3) (the observer inputs and outputs) between processors 1 and 3, and VKFILT (the filtered airspeed) between processors 2 and 3.

CHAPTER 14

UNPILOTED TIME SIMULATION

Typical manoeuvres were simulated on the non-linear model to check operation of the control law. On the basis of these tests, the only change required to the $\mathcal{H}_\infty$ feedback controller was a reduction in the bandwidths used for the nozzle and throttle because of the non-linear nature of the throttle-to-thrust characteristic, and backlash in the nozzle servo. However, the time simulations were very much part of the design loop when determining the gains and logic for the prioritized desaturation scheme. Setting these gains too high can be destabilising, whilst setting them too low doesn't give the desired priorities. The time simulations on the non-linear model do not allow for lateral motion of the aircraft. This is primarily because any lateral motion needs the pilot to be in the loop to stabilise it. Longitudinal performance in the presence of lateral motion was evaluated and checked during piloted simulation.

In the following sections the results of non-linear simulation of the final control law are presented. All the figures are located together at the end of the chapter.

14.1 Control law initialisation

Whenever the safety pilot is flying the aircraft manually it is ensured that the control law is always ready to come on-line. This is achieved by updating the observer states every cycle using the current plant inputs. The pitch channel off-line initialisation was slightly modified as discussed in chapter 11. The controller states are being initialised whenever IFLY = 2. Setting IFLY = 1 then brings the control law on line.

Although the software allows the control law to come on-line at any stage in flight, during flight testing the control law is only going to be engaged at a steady 250 knots at zero climb rate. Figure 14.1 illustrates bringing the control law on line. The variable CL005 is driven from IFLY and is set to 1 if the test pilot is flying, and 0 if the safety pilot is flying. Hence for the first 10 seconds the control law is off-line, and the airspeed and flight path start to wander off. When the control law comes on-line the flight path smoothly converges to zero and the airspeed settles to the demanded value. Note the the smooth response of the engine. Activity of the TNM and tailplane frequency limiter is not expected when bringing the system on-line in the aircraft.

14.2 Manoeuvre testing

Figure 14.2 shows the effect of speed demand changes in wingborne flight. At these airspeeds the two controlled variables are airspeed and flight path. The two speed changes show no significant coupling into flight path. Figure 14.3 shows flight path changes at a constant airspeed. For a 5 degree flight path change, the coupling into airspeed is about 5 knots which is minimal. To create the response in simulation the stick was set to a constant value, and then instantaneously set to zero when 5 degrees flight path angle was obtained. This results in an overshoot in the flight path, the size and speed of which is dictated by the bandwidth of the flight path.

Figures 14.4, 14.5 and 14.6 show manoeuvres in the hover initiated by the left and right hand inceptors and trim switch demands. INCEP1, INCEP2 and INCEP3 are throttle lever, stick and trim switch respectively. In all cases decoupling is very good, and should reduce pilot workload considerably. Even though height is not used for feedback, the height is maintained well using the height rate (VKD) feedback. Notice that the engine saturates during the vertical climb, and that good control is maintained despite this. The observer desaturation scheme is providing the anti-windup under these conditions.

14.3 Transition testing

Figure 14.7 shows a stick-free rapid deceleration from wing-borne flight to the hover. This is initiated by setting the left hand inceptor further than three-quarters the way back from centre position (or equivalently asking for a ground speed of less than -10 knots). The nozzles come forward immediately and the engine builds up to full power. Note that the nozzles are at their front-stops position for most of the manoeuvre, and that this is a good test of the desaturation scheme. The response is not perfect, but then flight path is not being controlled below 190 knots. The pilot can of course "fly-out" the flight path deviations. For a normal deceleration to hover, refer to the next section which deals with the effect of sensor noise.

Figure 14.8 shows a stick-free rapid acceleration from hover to wingborne flight. The manoeuvre is initiated by pushing the throttle lever fully forwards, leaving it there until the desired speed is reached, and then pulling it back to demand that speed. Flight path coupling is less than about 1.5 degrees. The pilot could easily fly this coupling out with small stick movements if he so desired.

14.4 Effect of sensor noise

Flight data from test of control law 003 was used to determine the sensitivity of 005 to sensor noise. The main concern is the tailplane frequency limiter which cuts out the control law if tailplane oscillations with frequency greater than 6 Hz and amplitude greater than 1% of full tailplane travel occur.

The underlying noise was extracted from the test data by filtering each sensor in turn with a low pass filter, and then taking the difference between the two signals to be the noise. The filter was a first order Butterworth for every case with cut-off frequency 10 radians/second.

Figure 14.9 shows a deceleration to hover with the noise signals added to the controller inputs. Performance is not noticeably affected at any flight condition. One second of the tailplane response is replotted in figure 14.10, and it can be seen that there is no discernible tailplane activity around 6 Hz.

14.5 Robustness evaluation

Sequences of tests for different fuel loadings and altitudes were carried out to provide confidence of the control law performance under varying conditions. Figure 14.11 shows a series of decelerations to the hover with total weights of 15000, 16500 and 18000 lbs. Figure 14.11 shows the earlier version of 005 with feedback around the nozzles over the 200–125 knot speed region. Figure 14.12 shows exactly the same test applied to the final version where a vertical thrust schedule is used over this speed region. Note that the original approach is much more robust in that the spread in VKD variation is less. However, the thrust schedule version had better dynamical qualities for the reasons discussed in section 8.3.

Figure 14.13 shows flight path responses between 1000 and 15000 feet. Figure 14.14 shows decelerations to the hover with model of the governor cut-in rpm varied (point D in figure 13.1).

Flight simulation is also very instrumental in checking the robustness of the design. Large lateral manoeuvres, and performance during acceleration correspond to off design point tests, and if the response is satisfactory give some degree of confidence in the design.

Testing using the original GVAM actuator models produces no discernible changes in the response. This gives some confidence in the robustness to actuator modelling. One effect not included in the GVAM model is the nozzle backlash. Figure 14.15 shows the the control law applied to the WEM model where the backlash is implemented. The presence of the backlash can be seen from the plots, but no instability or limit cycle results. An preliminary design with a higher bandwidth did initiate a limit cycle, and hence the bandwidth was reduced to its current value of 2 r/s.

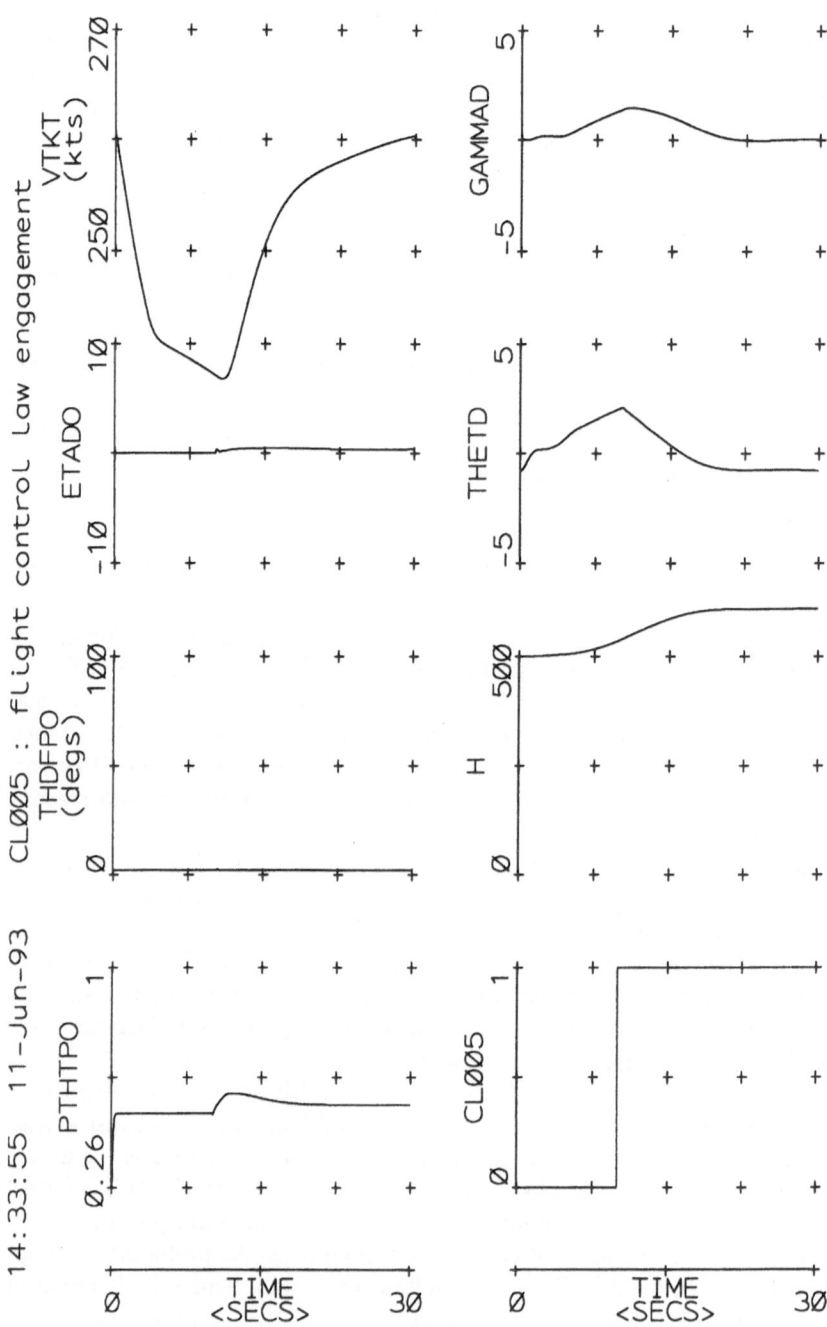

Fig. 14.1. Flight control law engagement

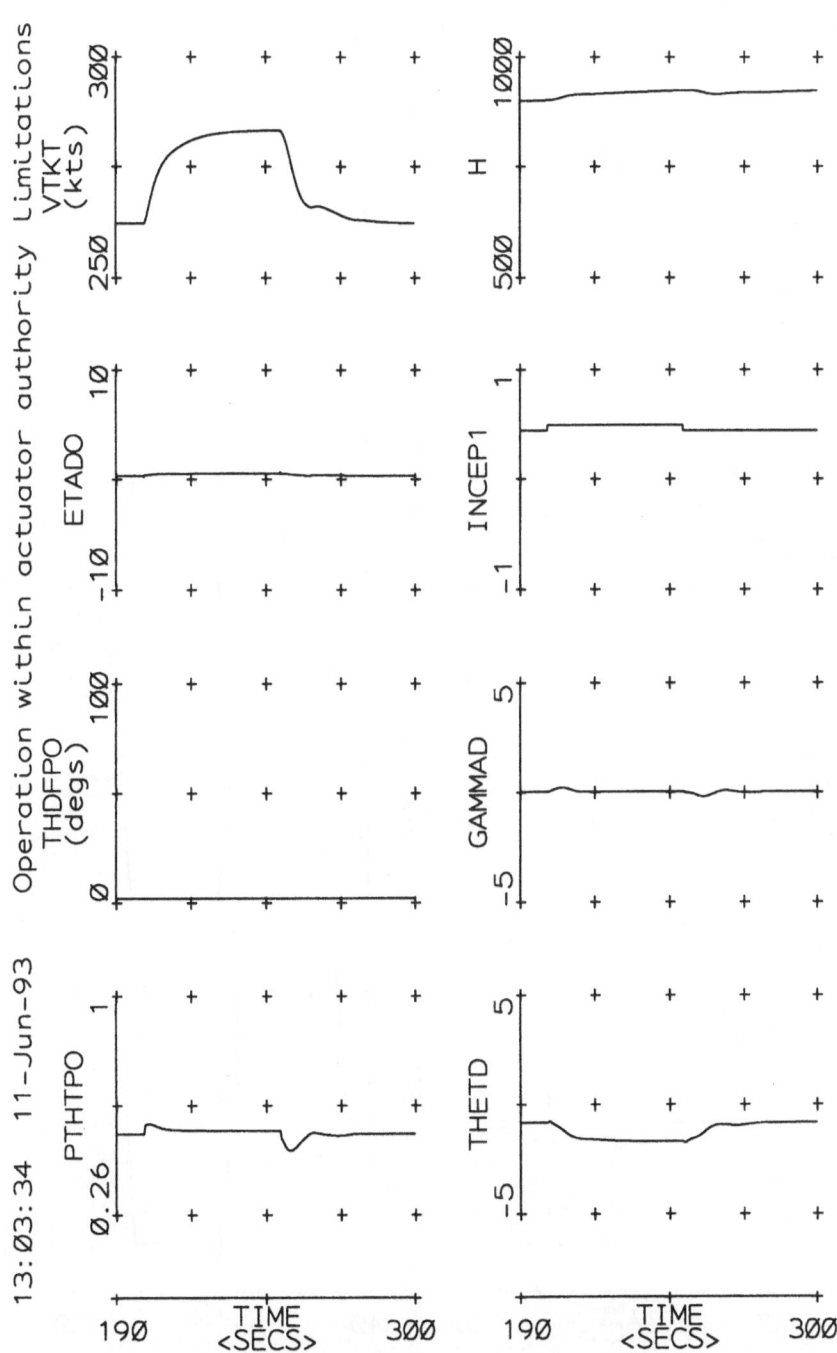

Fig. 14.2. Demanded speed changes in wingborne flight

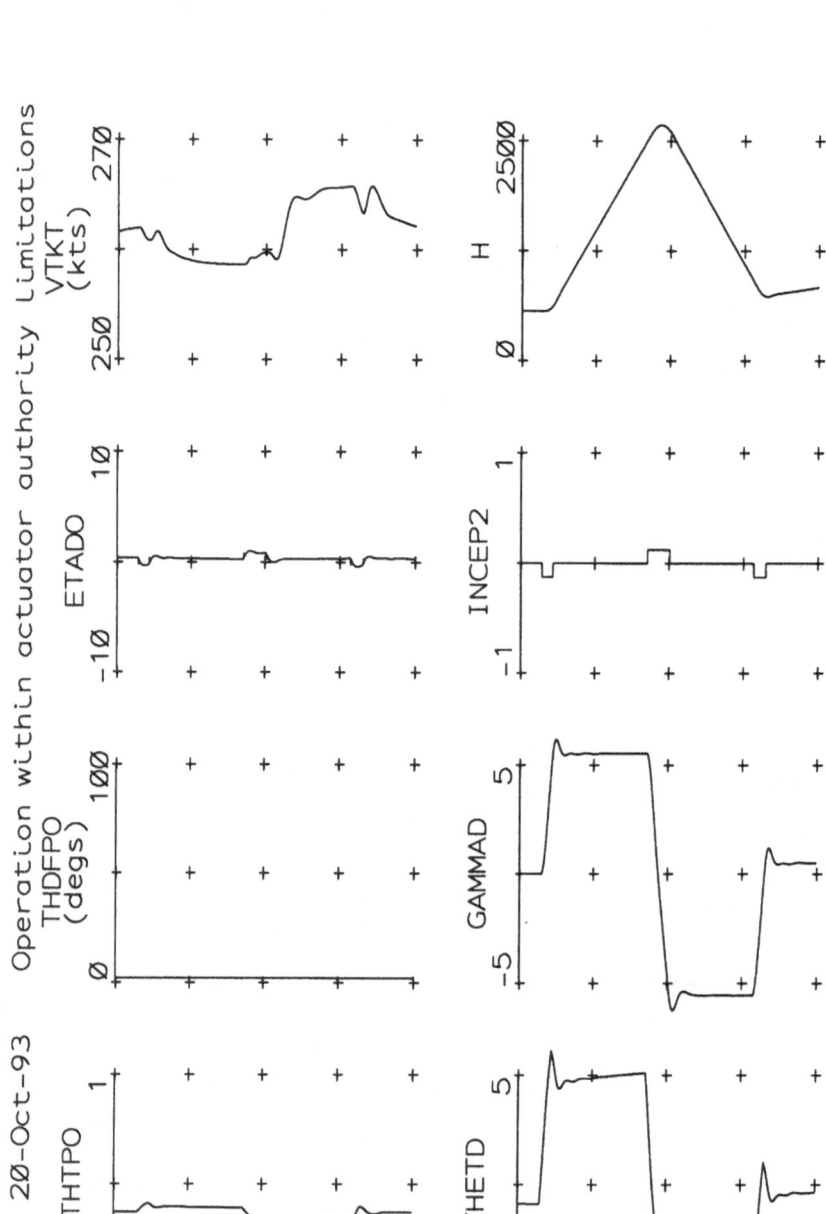

Fig. 14.3. Demanded flight path changes in wingborne flight

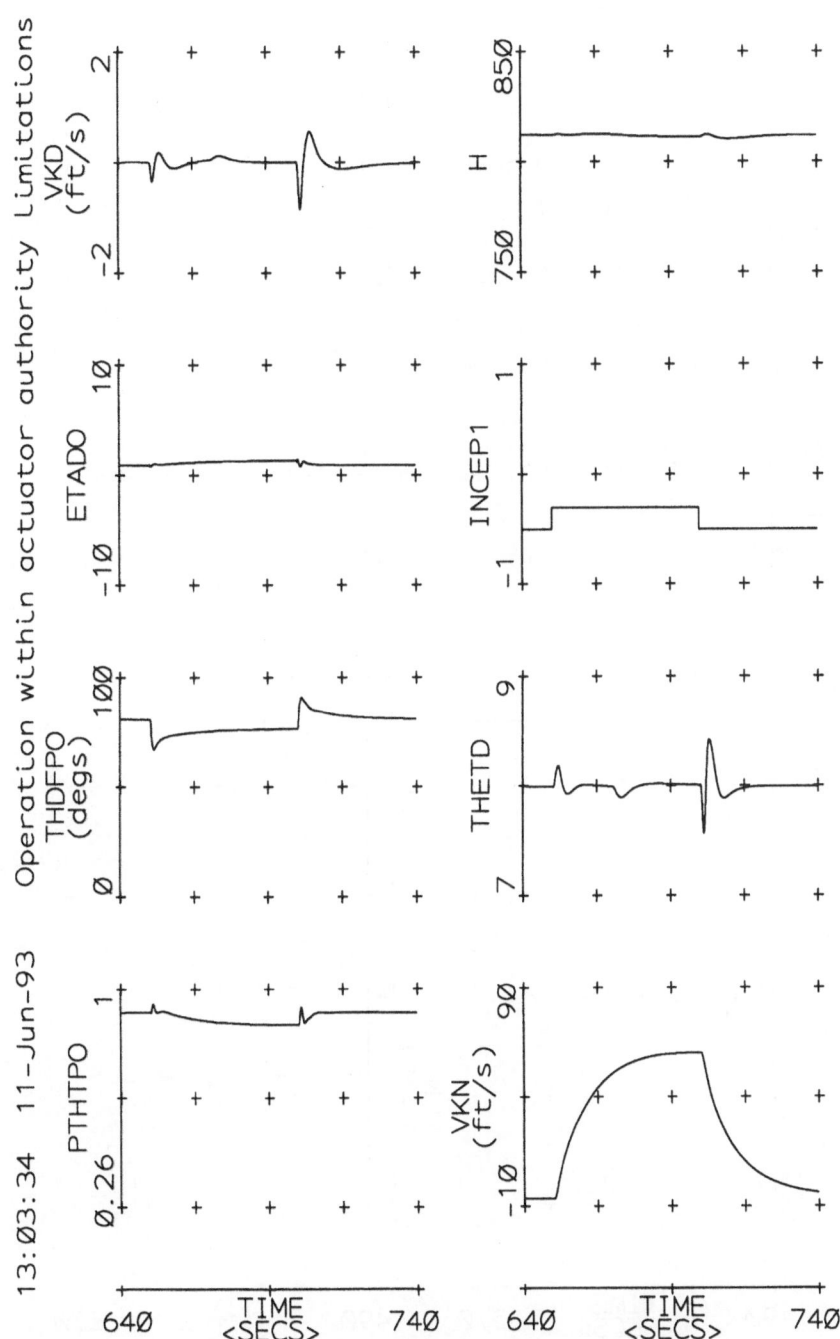

Fig. 14.4. Forward speed changes in the hover

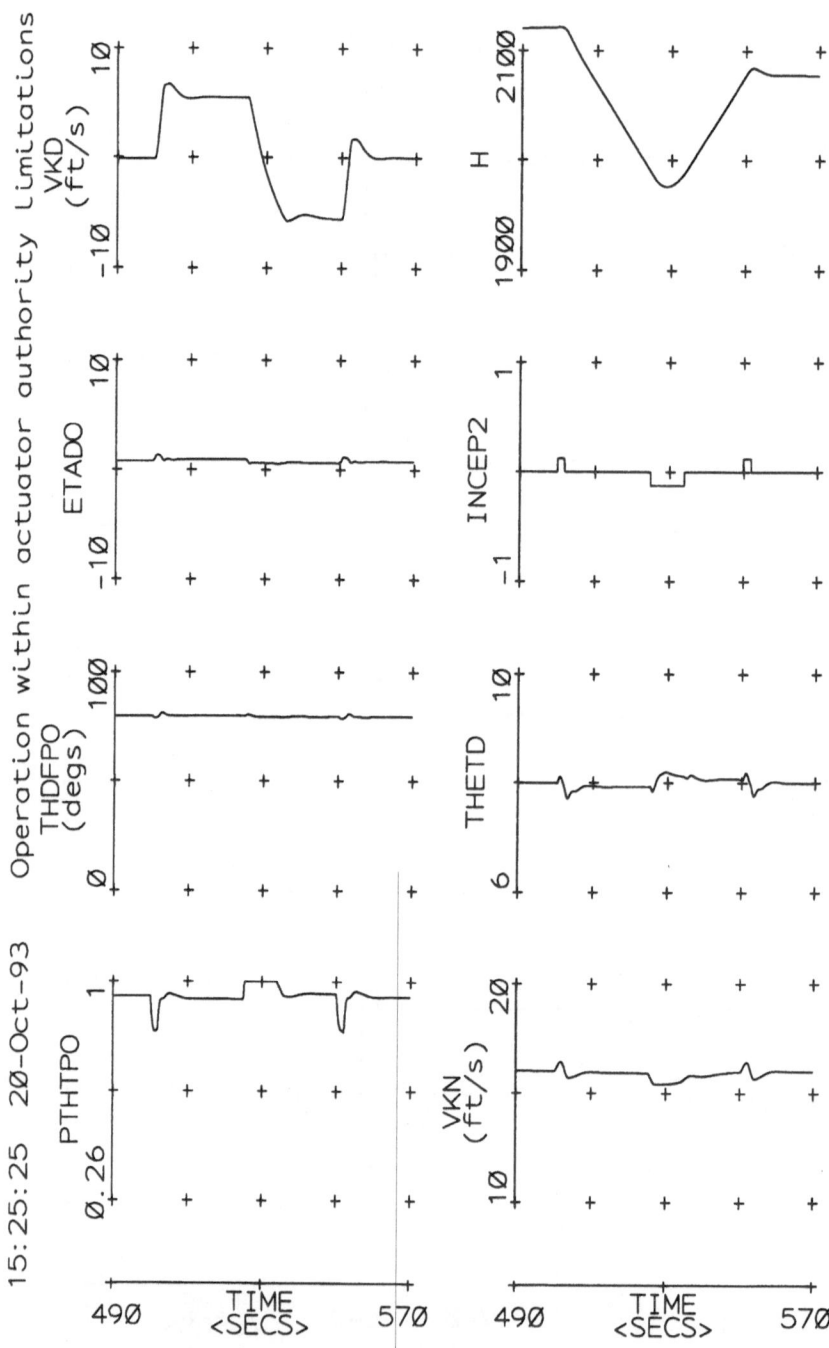

Fig. 14.5. Vertical speed changes in the hover

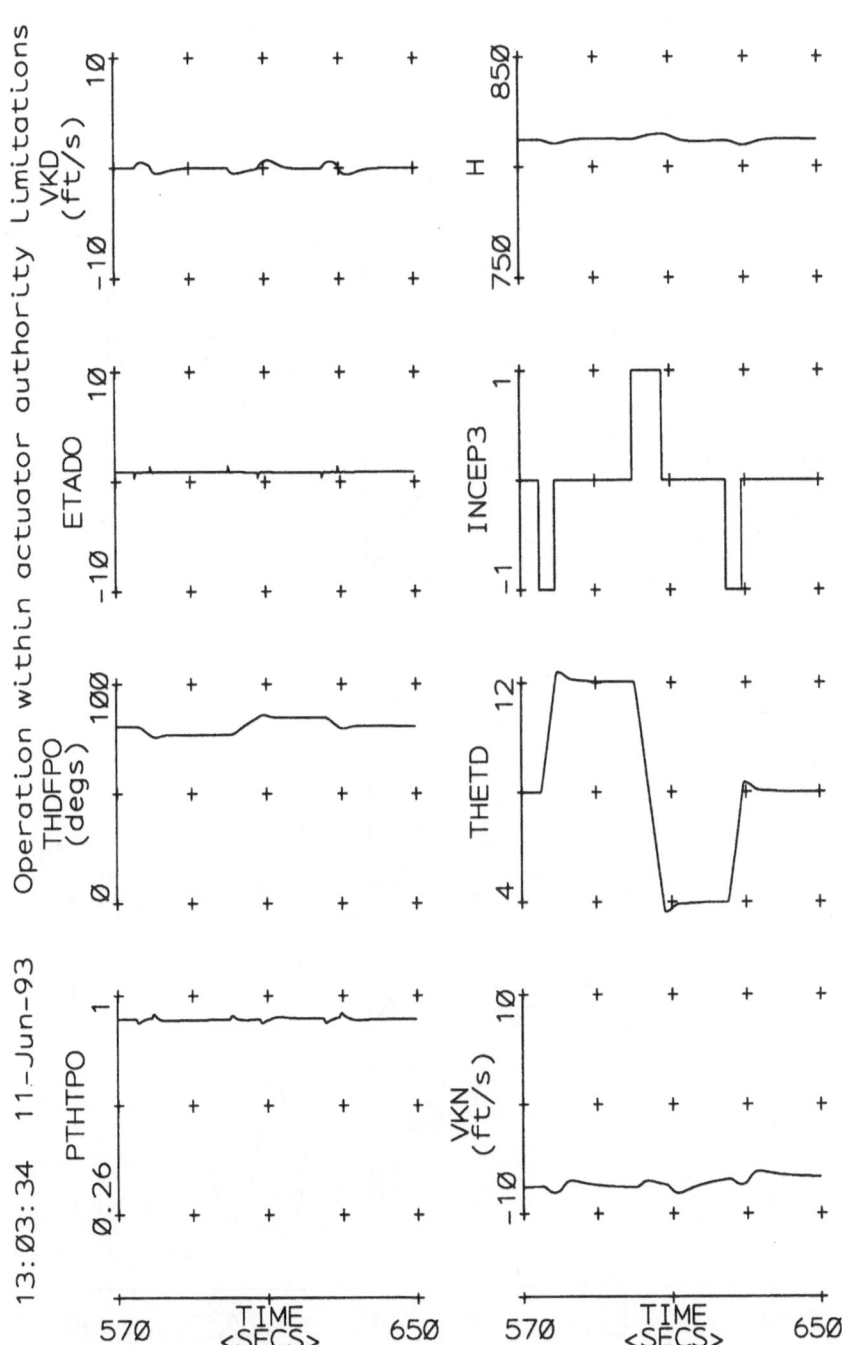

Fig. 14.6. Pitch attitude changes in the hover

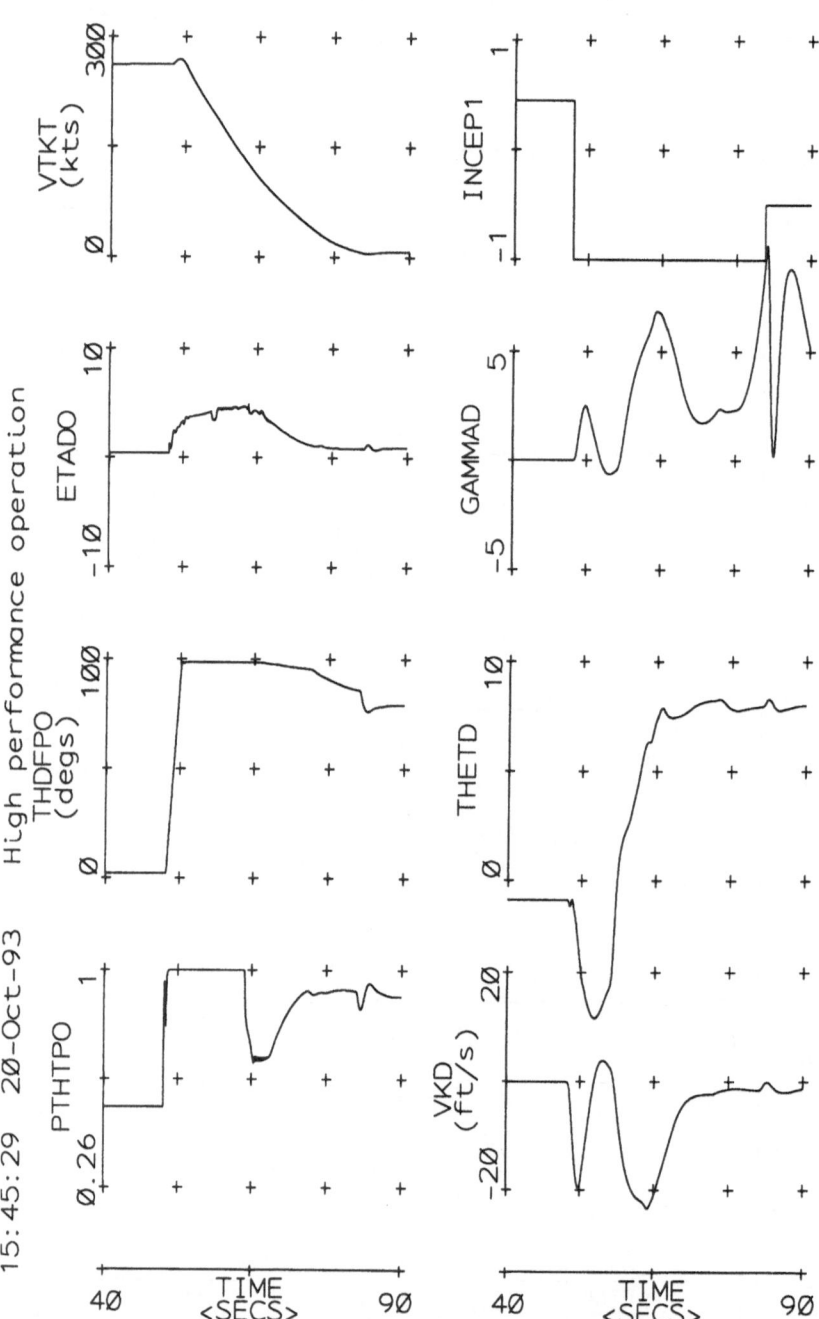

Fig. 14.7. Rapid deceleration to hover

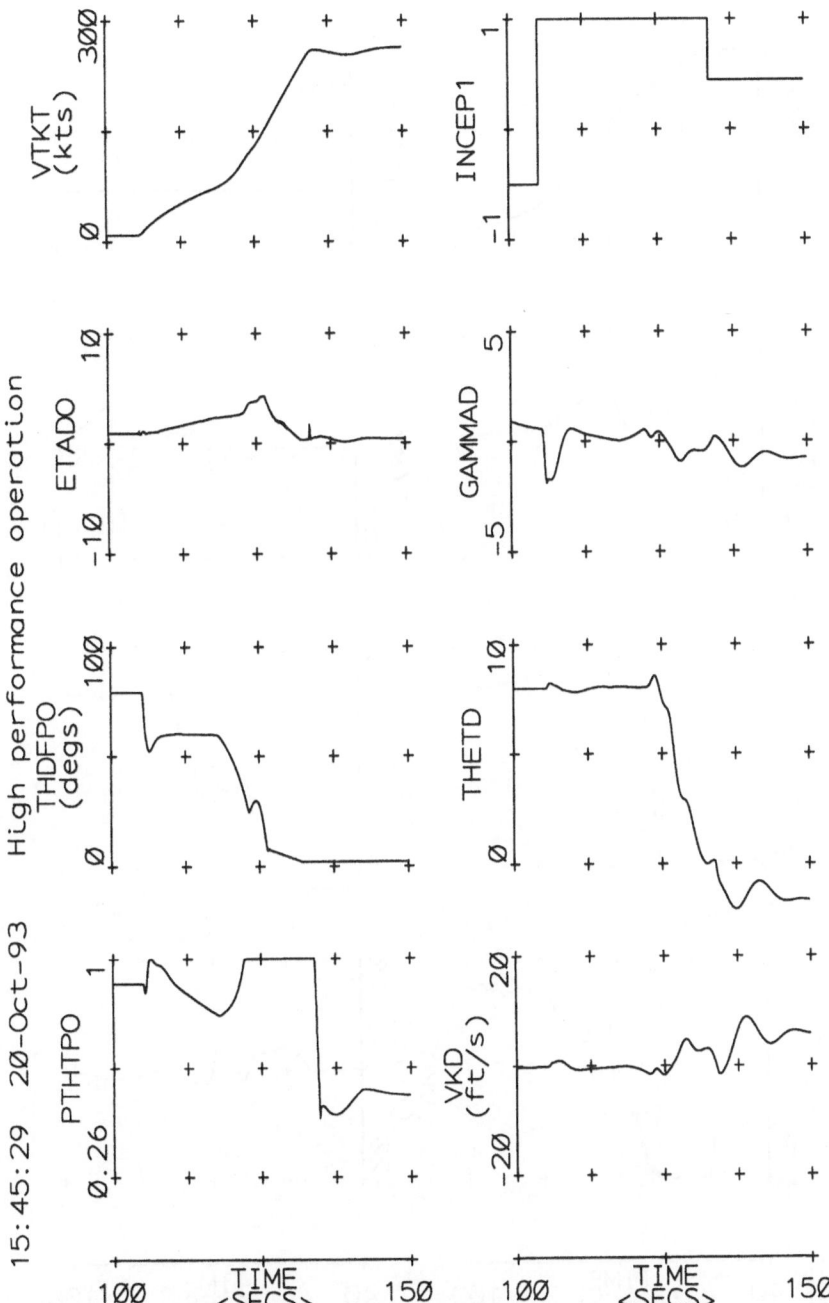

Fig. 14.8. Rapid acceleration to wingborne flight

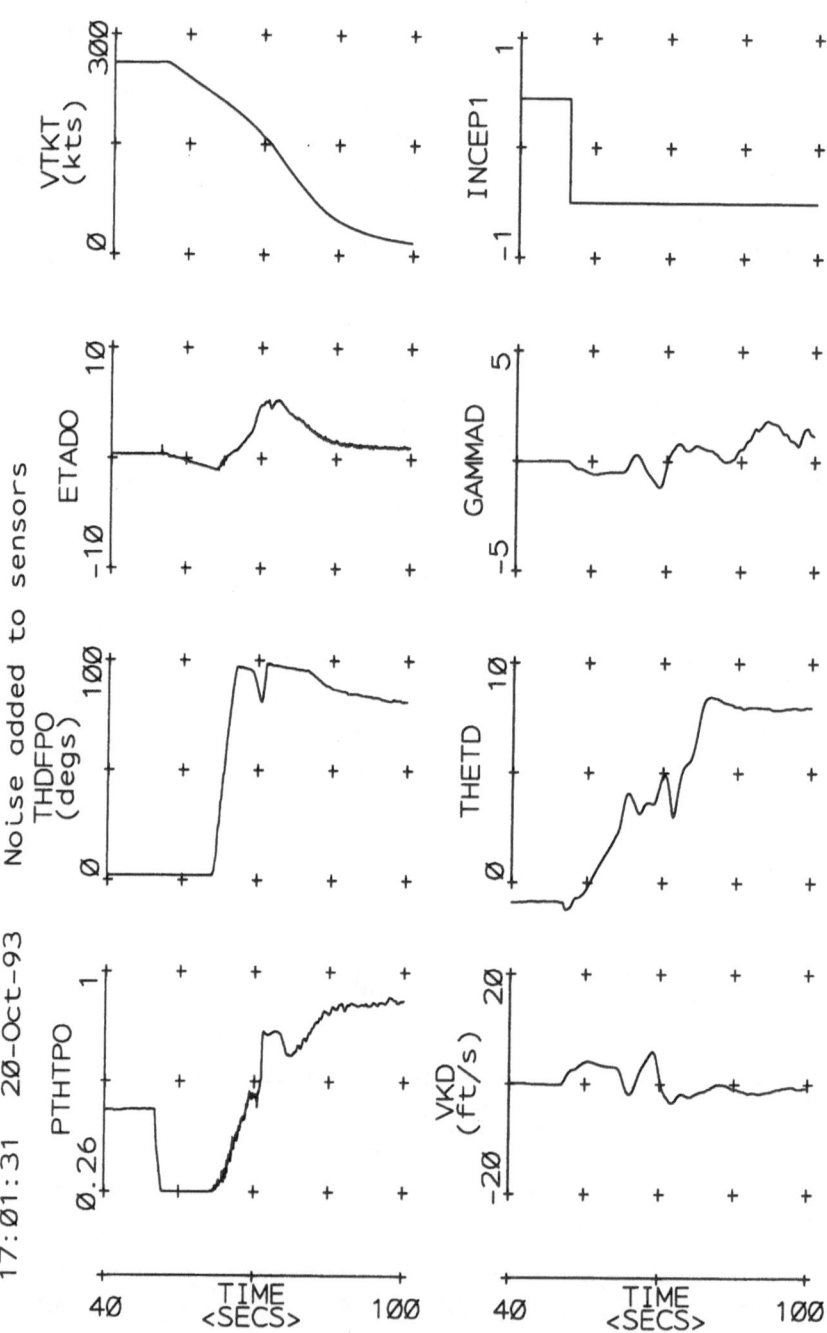

Fig. 14.9. Rapid deceleration to hover with sensor noise added

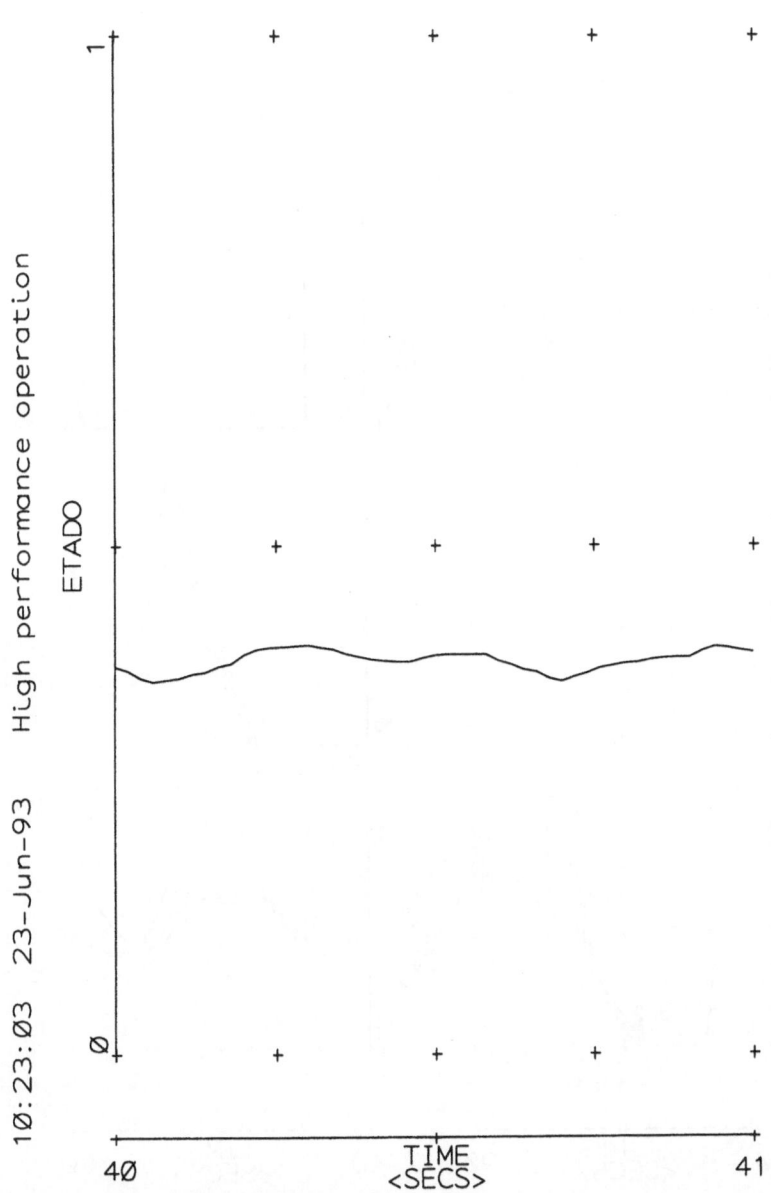

Fig. 14.10. Tailplane response sensor noise added

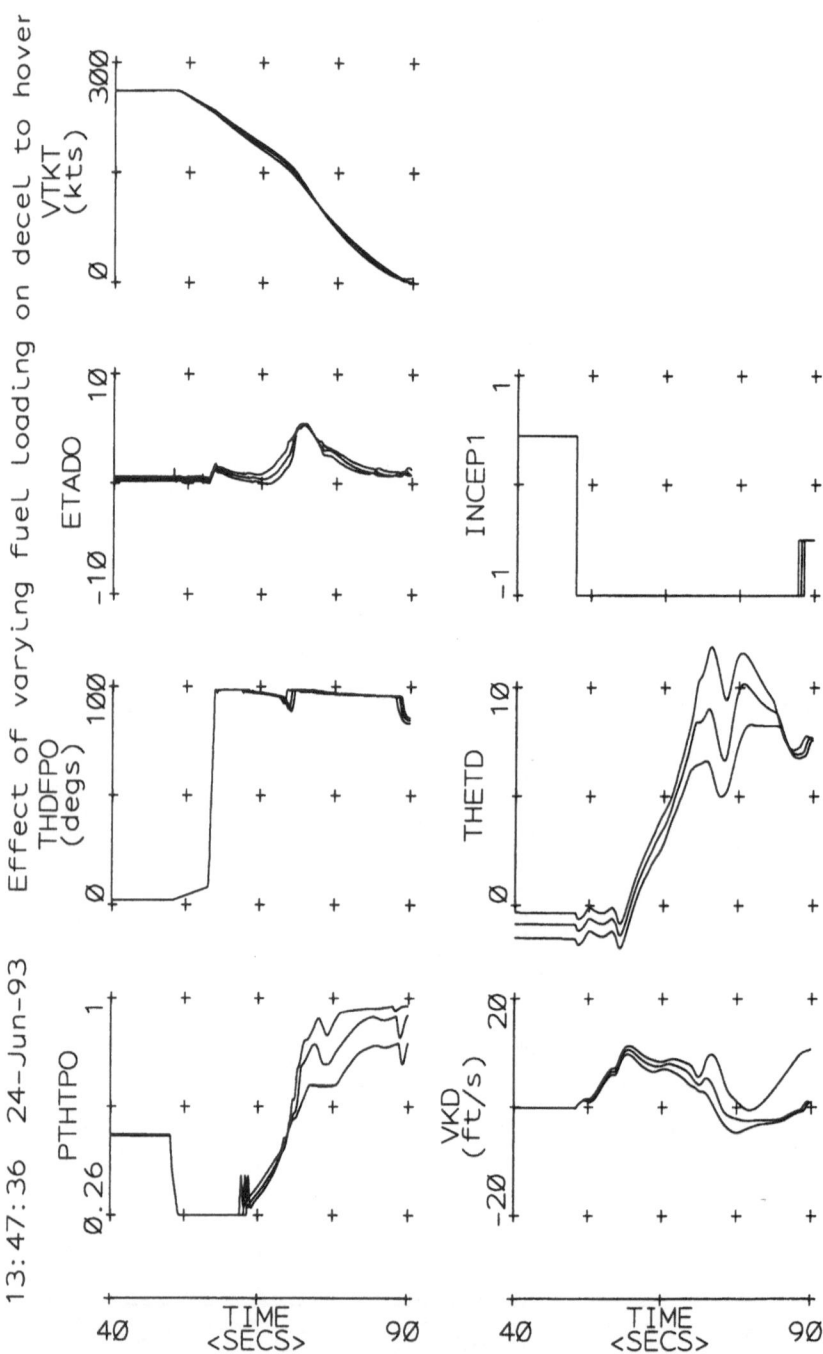

Fig. 14.11. Effect of varying fuel load on deceleration profile

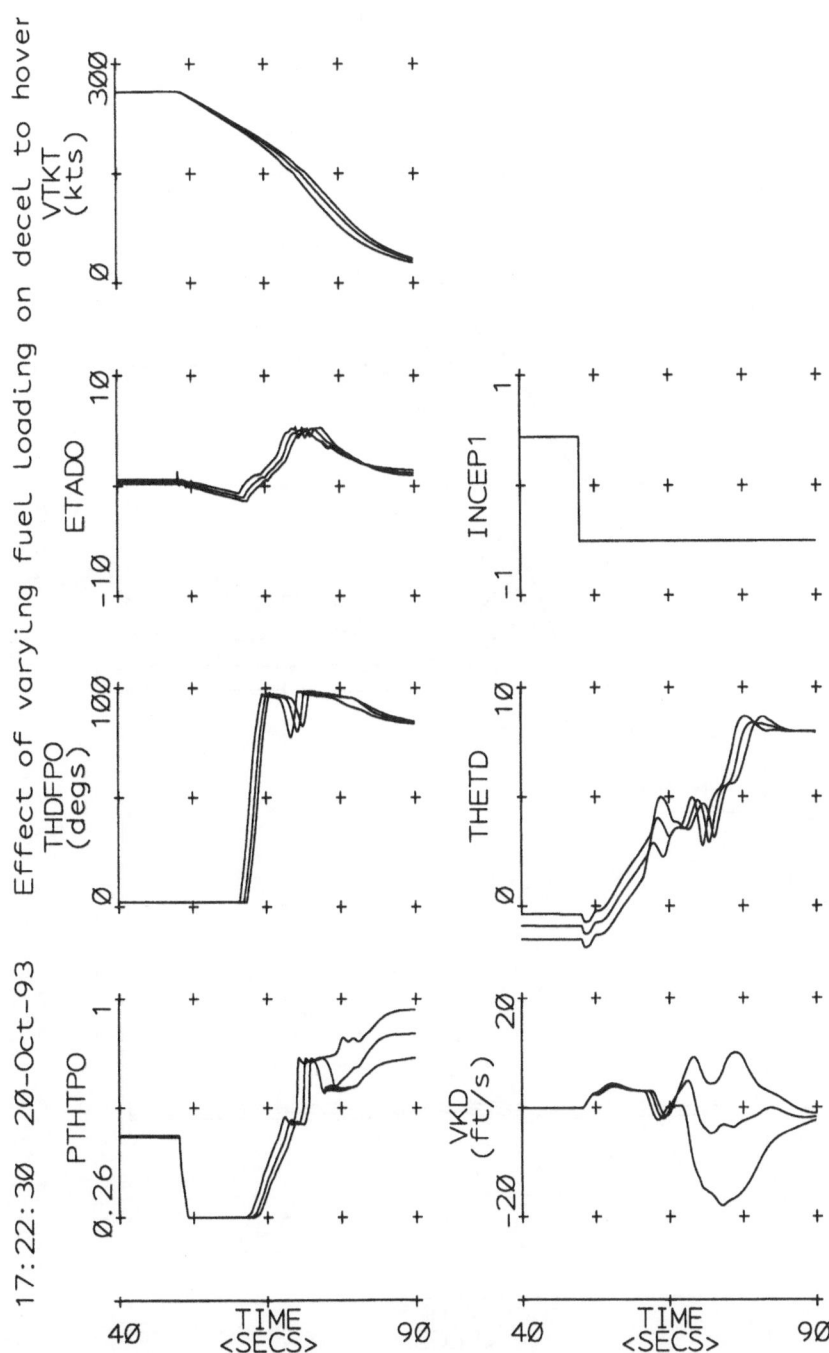

Fig. 14.12. Effect of varying fuel load on deceleration profile

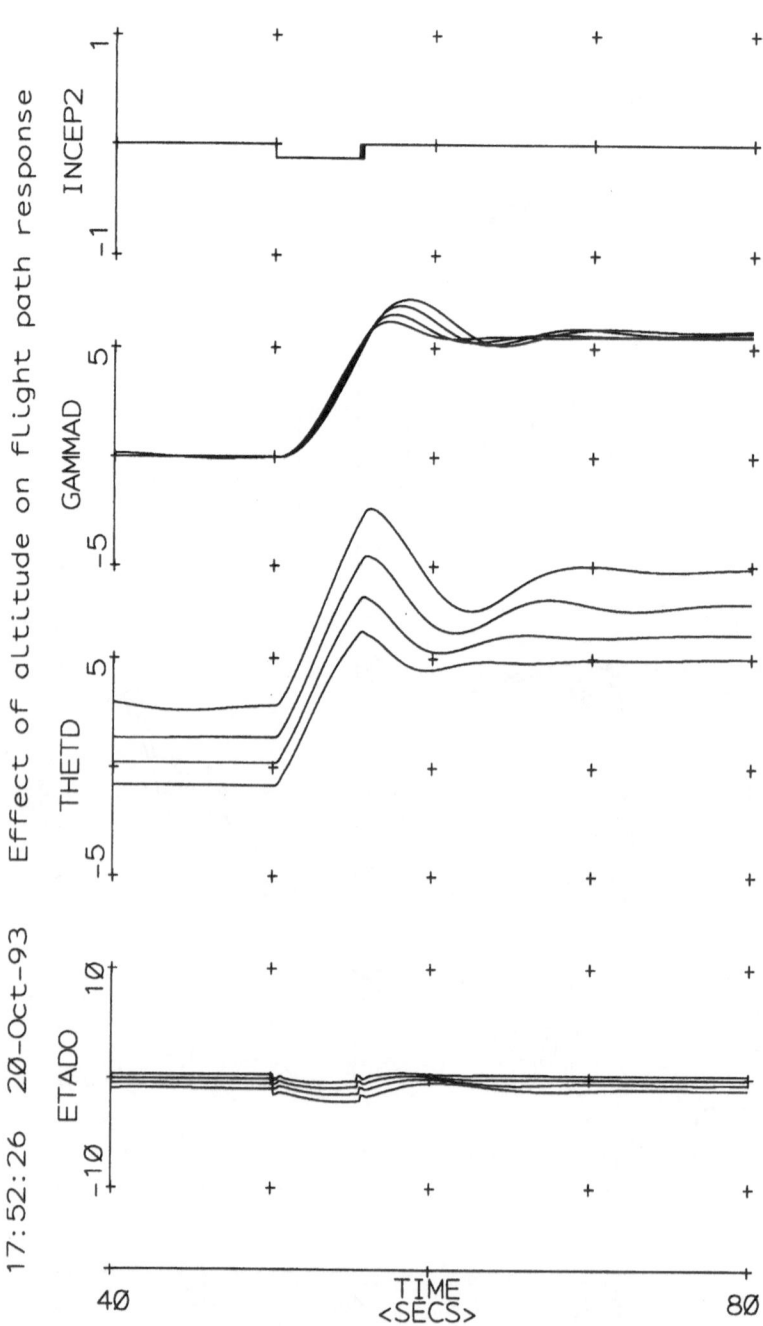

Fig. 14.13. Effect of varying altitude on flight path response

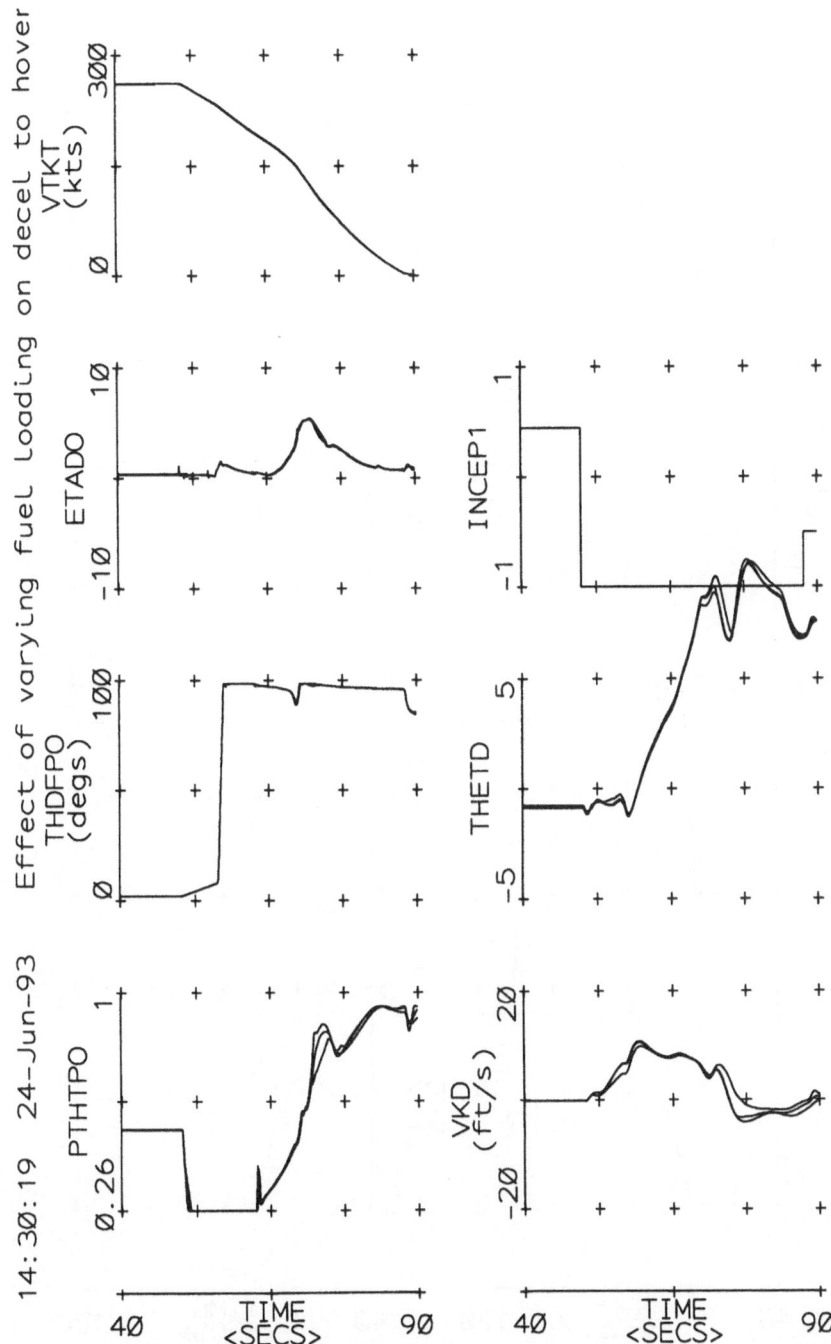

Fig. 14.14. Effect of varying ENJD on deceleration response

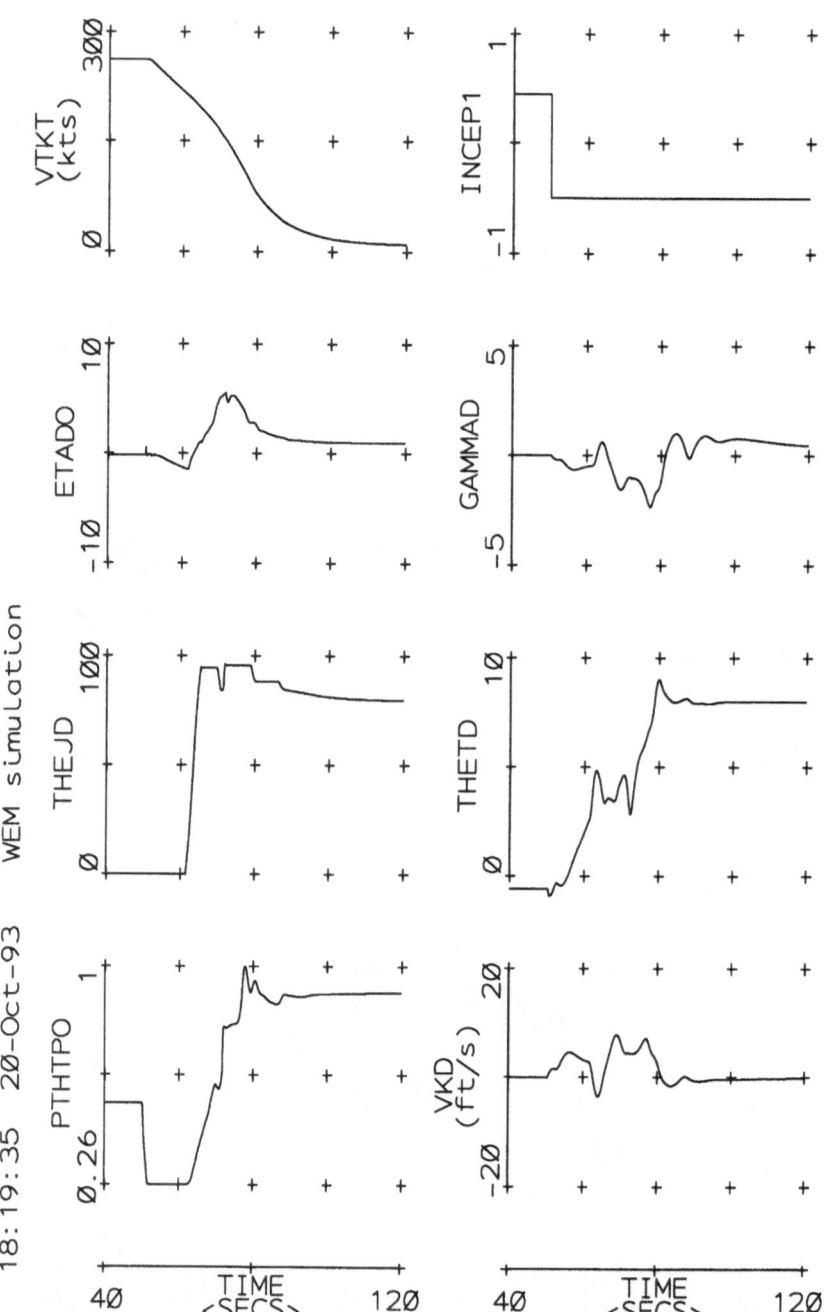

Fig. 14.15. Effect of nozzle backlash

CHAPTER 15
PILOTED SIMULATION AND FLIGHT TESTING

15.1 Simulator results

The control law went through several periods of flight simulation testing by both trained and un-trained pilots. Piloted simulation proved very important in the evolution of the control law, and showed up effects not discernible from non-piloted simulation. Some of these effects are as follows:

1. Pilot in the loop. When the pilot is flying the controlled aircraft he closes the loop between his inceptors and what he is attempting to control. For most tasks the pilots' attention is primarily directed towards controlling flight path angle (or height rate if in the hover) which is fed back to the pilot by the symbol in the centre of the head-up display. It is actually possible for the pilot to destabilise the system if the lag in the controlled system is too high. This effect is commonly referred to as pilot-induced oscillation (PIO).

2. Head-up display format. The form of the head-up display (HUD) and the information presented have a large impact on the assessment of the control law. For example, the original HUD used for jet-borne flight caused problems in that both pitch attitude and flight path are shown simultaneously. This meant that it was very easy for the pilot to pick up the wrong symbol when changing from the wing-borne flight HUD in a deceleration. The pitch attitude symbol was removed to resolve this problem, it being superfluous given that the control law sets the attitude for landing automatically. The precision to which variables are displayed was also found to be very important; if airspeed is displayed to the nearest knot, then there is a tendency to try and set the speed to this accuracy, even though this would not be normal practice.

3. Lateral manoeuvres. Lateral manoeuvres led to three compensation terms being added to the longitudinal control law. Feedforward was added from bank angle to airspeed demand in wing-borne flight to ensure no airspeed is lost. Feedforward was also added from bank angle to fuel flow demand when in the hover so as to prevent any height loss following a rapid bank demand. The prioritized desaturation scheme was also adjusted for engine saturation at large bank angles, backing off vertical speed demand if necessary.

4. Stick dynamics. The stick and hydraulic feel system have a relatively large lag which affected the perceived response of the control law. In particular the lack of a well defined centered stick position required a large dead-band to be implemented which could lead to difficult height rate capture in the hover.

5. Blend region. The speed region over which the blend between control modes is effected was altered several times. There was a trade-off between allowing the pilot to use aerodynamic lift down to as low a speed as possible, and ensuring that the aircraft stays within safe operating limits in terms of angle of attack and maximum descent rates on approach to hover.

6. Engine non-linearity. In the course of flight simulation a small engine limit cycle was found around the knee in the throttle–fanspeed characteristic. The size of this can be minimised by adjusting the model of this characteristic within the control law i.e. by adjusting ENJB and ENJD. However, the effect cannot be completely removed since the point at which the engine governor cuts in cannot be precisely predicted.

7. Flight path quickening. The HUD used for piloted simulation had flight path quickening. Whilst beneficial in terms of improving the perceived tracking performance with the pilot in the loop, it has a detrimental affect on the flight path hold facility. For instance, if the pilot tries to capture 5 degrees up flight path he will centre the stick when he sees the flight path symbol on the HUD reach 5 degrees up. With the flight path quickening the HUD symbol will be ahead of the actual flight path which in turn means the captured flight path will be some value less than 5 degrees. In piloted simulation the effect is that the flight path symbol drifts down to a lower value. A solution to this could be to use the quickened value of flight path to drive the flight path hold.

8. Speed hold authority. The authority of the speed hold is on the low side in wingborne flight. This is partly due to the limitations imposed by the engine non-linearity on bandwidth. An extra integrator might also help, but this was not done as the order of the controller would increase by 1, and the available processor power is insufficient to cope with this. Hence feedforward terms are used from bank angle to speed demand to help maintain airspeed in steep turns.

Pilots found flight path tracking to be satisfactory across the flight envelope. This is a result of having sufficient bandwidth from the feedback controller and the way the $\mathcal{H}_\infty$ controller has been implemented so as to minimise lag between references and outputs (see [4]). Performance could possibly be improved using a demand precompensator. In [36] a constant called the flight path time-delay is defined and identified as an important indicator of the ease with which flight path tracking can be effected. Figure 15.1 shows the evaluation of the flight path time delay for the control law at 220 knots. It is the intercept found by extending the asymptote of the flight path response to a step demand on the stick back to the x-axis. In [36] it is stated that a delay of 1.5 seconds is suitable for flight path tracking. Hence the flight path tracking task could be improved here. When no powered lift is being used, allowing pitch rate overshoot is the only way to achieve this. This could be added with a pilot demand precompensator. The concept of flight path time delay can be extended to the jet-borne lift regime, and this results in delays of approximately 0.5 seconds for airspeeds below 105 knots indicating good handling qualities. In [36] guidelines as to when PIO is likely

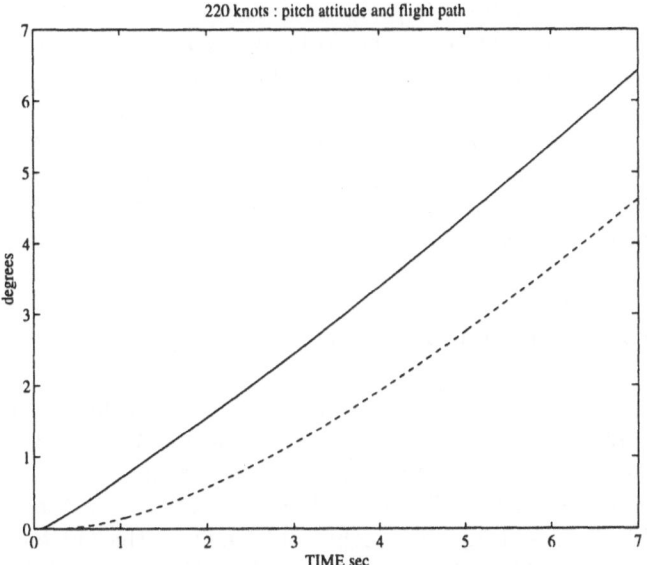

Fig. 15.1. Pitch attitude (solid) and flight path (dashed) following a step demand on the stick at 220 knots

are also given. The two key parameters are identified as

1. the frequency at which the transfer function from pilot demand to controlled output reaches 180 degrees, and
2. the rate of increase of phase lag with frequency at this frequency.

Table 15.1 illustrates these parameters for the control law in wing-borne and jet-borne regimes.

Table 15.1 – PIO susceptibility			
Airspeed (knots)	Controlled variable	ω_{180} (rad/s)	Phase rate (^{o}/decade)
65	Height rate	2.5	125
200	Flight path	1.1	125
200	Pitch attitude	5.0	150

The 180° frequencies for flight path and height rate control are on the low side, even though no PIO problems were encountered when flying these tasks. Pitch pointing, however, should be very good. The phase rates could also be improved in all cases with some phase advance to bring them closer to the 90° per decade identified as the ideal case in [36]. One pilot did get into PIO when trying to capture height. This task adds an extra 90° phase lag relative to the height rate capture, and so the PIO is not surprising. A button which demands height hold when pressed was added to the control law to improve the height capture task.

Results from data logging

Various manoeuvres were tested on the final version of the control law, and the results of some of these are presented here. Figure 15.2 shows a climb in the hover from 500 to 600 feet and capturing 600 feet without the help of the height hold facility. Figure 15.3 shows a descent in the hover along with height capture using the height hold facility - the button which sets HENABL is pressed at 400 feet. Figure 15.4 shows the aircraft being yawed round by 90 degrees and with no longitudinal stick inputs from the pilot. The control law maintains height reasonably well - the initial loss of height and final gain in height are the result of engine bleed air being used to control the yaw angle. Figure 15.5 shows banking in the hover to set up a lateral velocity. Note that height is maintained at the demanded level. Finally figure 15.6 shows a deceleration to hover.

15.2 Flight test results

Flight control law 005 was first tested on 9th December 1993. It engaged first time without significant transients. For the control law to be engaged, it must be demanding actuator positions within ten percent of their actual positions. The observer/Hanus off-line conditioning used ensures that the demands are close, and it was commented that the control law engages nice and quickly.

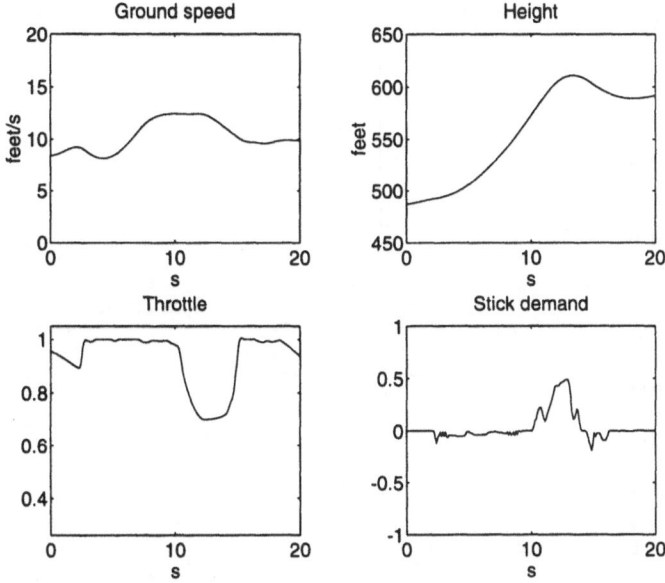

Fig. 15.2. Height capture using vertical velocity demand

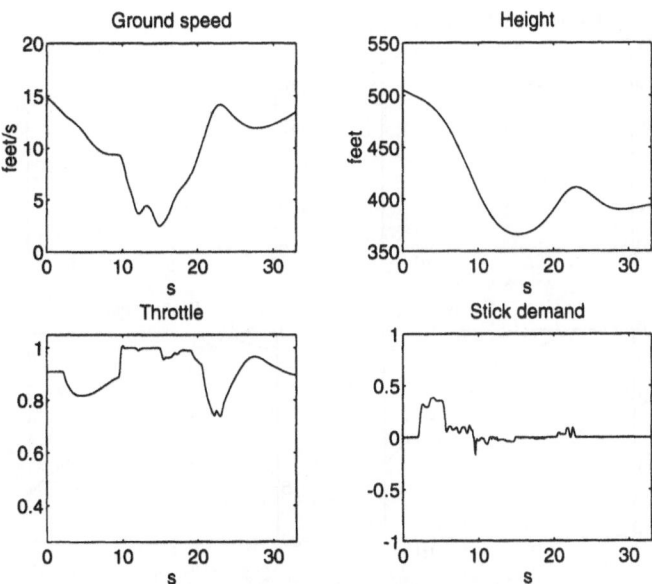

Fig. 15.3. Height capture using the height hold facility

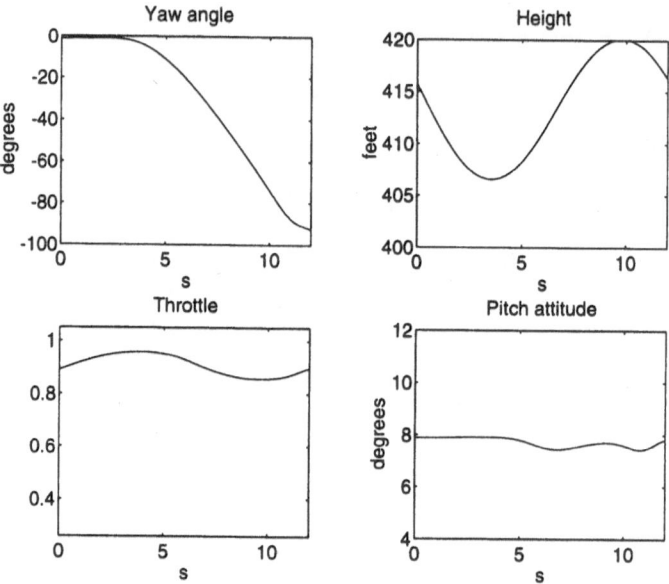

Fig. 15.4. Yaw manoeuvre in the hover

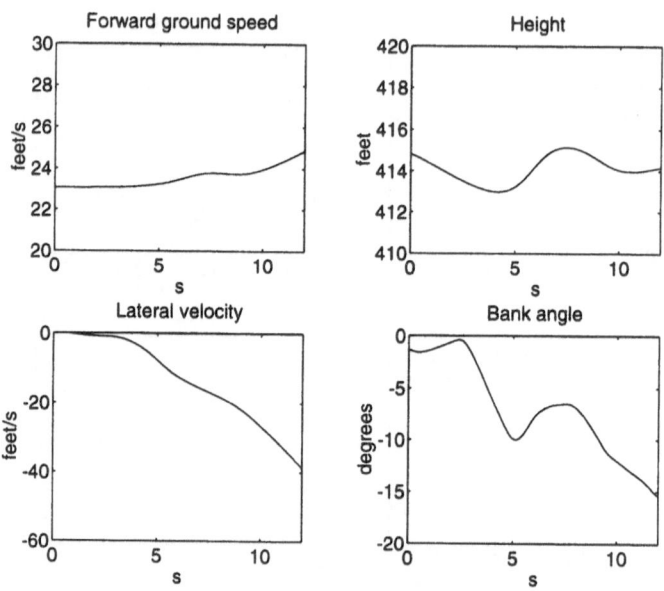

Fig. 15.5. Banking manoeuvre in the hover

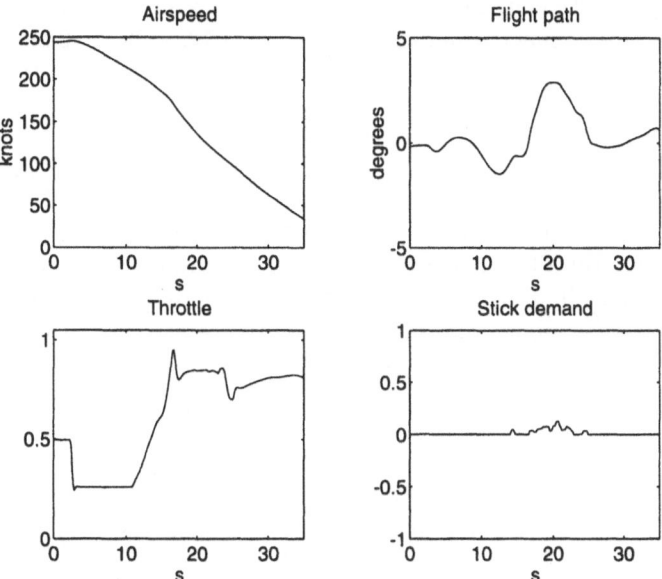

Fig. 15.6. Deceleration to hover

For the first part of the flight no inertial reference signals (IRS) were used by the control law i.e. only aerodynamic plus signals derived from accelerometers and gyros where used. These signals are inherently more noisy than those generated by the IRS. The control law coped well with both sets of signals, but as anticipated performance was much smoother using IRS data. Flight path changes were achieved accurately and with ease by the pilot, and flight path wander after releasing the stick was approximately a quarter of a degree..

There was no sign of the onset of instabilities during the flight. This is testament to the control law having sufficient robustness to any differences between the design model and the aircraft. The confidence gained from this would suggest that in future flights the bandwidth of the control law could be increased. However, given the current limit of 40 milliseconds for the sampling period imposed by the computational load of the control law, increasing the bandwidth may not be possible.

In a banked turn, the feedforward from bank angle to airspeed demand was too aggressive and halved from the value used on the simulator. The non-linear compensation for the engine governor occasionally caused rapid throttle movements in the 200 knot airspeed region. Given a signal from the engine which indicates when the governor is active, this effect could easily be removed.

On the second flight, the control law was tested all the way down to the hover. The main adverse comment was that the 125–105 knot blend region is too narrow in that it results in a relatively fast pitch up to the landing attitude of 8 degrees. The pitch trim response was particularly liked, and general hover manoeuvering presented no problems.

Following the first flight tests, no modifications to the H- infinity controller are proposed. All modifications will be external to the loop or concerned with the blend variables. For first flight trials, the tests carried out went very smoothly. This may in part be seen as a result of the coprime factor uncertainty modelling being appropriate for the actual uncertainty encountered between the model and the aircraft.

CHAPTER 16

CONCLUSIONS

The development and flight testing of this control law have demonstrated the practicality of $\mathcal{H}_\infty$ techniques. Aerospace applications may particularly benefit from this kind of design approach because of their high coupling and inherent multivariable nature. In the process of the design, many issues concerned with design and implementation of both multivariable and H-infinity optimal controllers have been addressed. The result is an exremely powerful overall design methodology which could be applied to a multitude of other problems.

The potential advantages and drawbacks of the control law design strategy used for 005 are now discussed. The word "potential" is used to emphasise the fact that comparisons with other design methods are not always clear-cut or appropriate. The main purpose of developing 005 was to demonstrate the viability of $\mathcal{H}_\infty$ and multivariable design techniques for application to flight control. However, having made a note of caution, it is clear that the initial flight tests of the control law were very successful. In particular the whole flight envelope was tried out in the first two flights. The handling qualities of the control law could no doubt be improved in future designs by increasing bandwidths as confidence in the model is gained and also by improving the pilot command precompensators. At present performance is severely limited by the processor power available which necessitates a slow sampling rate. It is anticipated that in the future this will present less of an issue.

Advantages

Taking a multivariable approach results in good performance and robustness in the hover. Both the engine and nozzle have similar maximum bandwidths, and the multivariable approach allows both to be used to their full potential. One would expect that as a result better wind gust rejection would be possible than if successive loop closing were done. With a given set of bandwidths at which the three actuators are to be used, robustness is also likely to be better.

Using feedback around the nozzles during transition (125 through 200 knots) may potentially give a more predictable response than if a vertical thrust schedule is used. Although the final version of 005 used a thrust schedule, it is believed that the feedback solution can be made to work. The question of how to use this extra degree of freedom effectively still needs to be answered.

Assuming that a multivariable approach is to be used, then there are then certain advantages to using the particular multivariable design approach used here. These are now discussed:

1. We have found the $\mathcal{H}_\infty$/loop-shaping linear design approach easy to use, and to produce very robust controllers with a modest amount of effort on the part of the designer. In particular, initial design studies very quickly reveal what is achievable from a given plant. Also, given an initial design, it is usually very clear how to alter the weights in order to achieve a desired result e.g. less use of a particular actuator, or tighter control of a particular output. In a sense the approach should be seen as a design tool as opposed to an automatic process which produces a controller.

2. The observer structure has been central to the way the control law was implemented. It results in good step responses to closed-loop demands (little or no overshoot and low phase lag), and has been used for the scheduling and anti-windup schemes. This is seen as a very strong advantage for this particular $\mathcal{H}_\infty$ design formulation, though it is possible that a similar approach could be used for $\mathcal{H}_\infty$ controllers in general.

3. Design time for the overall control law was also relatively short. One of the reasons for this may be that using feedback as opposed to operating point look-up tables is a less time-consuming task. Re-design time for a modified aircraft e.g. for an altered wing area, would be expected to be very easy, and could possibly be partially automated.

4. The advantages of taking a multivariable approach may be much more pronounced for future aircraft where coupling is likely to be much higher, and may even necessitate the use of multivariable design.

Limitations

The control law requires roughly twice as much processor power in comparison other control laws evaluated on the VAAC Harrier. This is due to the relatively large dynamic order of the controller which results in the running of the state equations being computationally intensive. This necessitated doubling the sampling period to 40 milliseconds which in turn limited the bandwidth which could be used for the pitch loop. However, given the trends in computing power over the last decade, this may not be a serious drawback in the future.

The study in chapter 8 investigating direct parameter optimisation indicated that large reductions in controller order might be possible for modest compromises in performance. Present model reduction techniques were not able to make this trade-off between order and performance for large state reductions. However, they did work well when removing states corresponding to relatively small Hankel singular values.

The prioritized desaturation scheme is the weakest link in the design approach in that its design is highly iterative. However, use of the observer desaturation combined with command conditioning proved very successful, and minimised the need for the prioritized scheme.

APPENDIX A
LIST OF VARIABLES

AETAD	tailplane/reaction servo demand
AFTRES(3)	aligned closed-loop error
AINVER	disables pitch and flight path holds
ALFAD	angle of incidence (degrees)
ANTIWU(10)	prioritized desaturation gains
APF	throttle servo demand from the linear controller
APFL	rate limited APF
APTHTP	throttle servo demand
ATHDFP	nozzle servo demand
AW1(8,8)	A-matrix for dynamic precompensator
AXCG	body axis forward acc. including component due to gravity
AXCGB	body axis forward acceleration (normalised 'g')
AXCGE	Earth axis forward acceleration (normalised 'g')
AXCGFB	Longitudinal acceleration signal for feedback
AXF0	longitudinal thrust demand
AXF1	authority limited AXF0
AXFL	achieved longitudinal thrust component
AXFLGR	achieved longitudinal thrust component relative to E-axes
AYCGB	lateral acceleration signal for feedback
AZCG	body axis vertical acc. including component due to gravity
AZCGB	body axis vertical acceleration (normalised 'g')
AZCGE	Earth axis vertical acceleration (normalised 'g')
AZCGFB	normal acceleration signal for feedback
AZF0	normal thrust demand
AZF1	authority limited normal thrust demand
AZFL	achieved normal thrust demand
BLENDE	blends between the two engine controller regimes
BLENDG	blends from vertical 'g' to 'q' mode
BLENDN	blends between low throttle desaturation scheme priorities
BLENDP	blends from 2ip-2op control to 3ip-3op control
BLENDR	blends in the lateral manoeuvre desaturation terms
BLENDW	60-100 knot resolved thrust to airspeed blend
BLENDX	speed authority demand limiting
BLENDY	speed authority demand limiting

BNKAUT	bank to fuel demand authority
BNKFIX	bank to airspeed demand authority
BW1(8,3)	B-matrix for dynamic precompensator
CONT	pitch rate demand from the trim switch (degrees/sec)
CONX	dynamically compensated left-hand inceptor demand
CONZ	stick demand after the addition of the dead-band
CTRLO(3)	Output of the state feedback matrix
CTRLU(6)	Input vector to the observer
CTRLX(20)	Observer state vector
CTRLX1(20)	Dummy vector used to calculate CTRLX(20)
CTRLX2(20)	Dummy vector used to calculate CTRLX(20)
CW1(3,8)	C-matrix of the dynamic precompensator
DBSTC1	current value of the stick dead-band
DBSTCK	stick dead-band nominal value
DAN1	if set to 1.0 then stick dead-band non-constant
DEMP	closed-loop pitch attitude demand (degrees)
DEMPT	pitch attitude hold demand
DEMSPD	signal to drive speed index on HUD
DEMX	closed-loop ground speed demand for hover/ transition modes (knots)
DEMXDM	demanded speed in closed-loop units
DEMXG	closed-loop airspeed demand for high-speed mode (knots)
DEMZ	closed-loop vertical speed demand for hover transition modes (knots)
DEMZG	closed-loop pitch rate demand for high-speed mode ($rads^{-1}$)
DEMZX	integrator state for the stick demand
DESATU(4)	prioritized desaturation demands
DISTI1(3)	disturbance on the input to the shaped plant
DISTI2(3)	disturbance on the output of the shaped plant
DISTO1(3)	feedback term to the plant input
DISTO2(3)	plant output + output disturbance
DREF(3)	scaled closed-loop demand
DRLMAU	descent rate limit (feet/minute)
DRLMCL	descent rate limit (closed-loop units)
DVLOAD	used to divide the scheduling load over several cycles
DW1(6)	D-matrix for the dynamic precompensator
ENABLZ	disconnects the stick integrator during engine saturation
ENGGOV	indicates when the engine governor is active
ENJA	engine x-axis intercept
ENJB	fuel flow demand corresponding to governor cut-in point
ENJC	fuel flow demand corresponding to max fan speed
ENJD	fan speed at which the governor cuts in

ENJE	maximum fan speed
ENJX(6)	coefficients for the dynamic engine compensator
ERRCL(3)	closed-loop error
ETACP	cockpit stick demand / test pilot's stick demand
ETADA	tailplane servo demand
ETADAQ	rate limited control law tailplane demand
ETADX	control law tailplane demand
ETATRM	trim-switch demand
FFPAUT	authority of the thrust/nozzle to tailplane feedforward
FFTH2P	tailplane feedforward term
FGAIN	pitch complementary filter gain
FILTPD	filtered pitch attitude signal
FMX(2)	coefficients for the measurement filters
FNP	engine fan speed
FNPL	engine fan speed
FPS2NG	scaling factor from feet/s to units of 'g'$\times$s
GAMDEM	flight path hold facility demand
GAMMAD	flight path angle (degrees)
GOVACT	set to 1 if the governor is active, 0 if not
H	altitude (feet)
HACTIV	height hold active flag
HDEM	height hold demand (feet)
HDOT	radar altitude derived climb rate (feet/s)
HENABL	height hold enable switch
IFLY	control flag : 0=desk, 1=cockpit/on-line, 2=manual/off-line
JDCL	set to 1.0 if just decelerated to hover
OFFST1	offset for the cockpit throttle lever
OFFST2	offset for the cockpit stick
OFFST3	offset for the desk left-hand inceptor
OFFST4	offset for the desk right-hand inceptor
PITCHF	pitch loop feedback variable
PLAM	λ_p : pitch attitude to pitch rate feedback ratio
PTHT11	dummy variable used in the engine control law
PTHT22	dummy variable used in the engine control law
PTHTPL	software generated low limit on throttle
PTHTPQ	throttle servo demand prior to rate-limiting
PTHTPX	control law throttle servo demand
Q,q	aircraft pitch rate (radians/sec)
QD	aircraft pitch rate (degrees/sec)
REF(3)	un-scaled closed-loop demands
RTENOZ	nozzle rate limit (degrees/second)
RTETHR	throttle servo rate limit
RTETLE	tailplane rate limit (degrees/second)

SCHDLE	vertical thrust schedule
SCLALF	alters alpha/flight path relationship in transition
SCLLE1	scale for cockpit throttle lever demand
SCLLE2	scale for cockpit stick demand
SCLLE3	scale for desk left-hand inceptor demand
SCLLE4	scale for desk right-hand inceptor demand
SCLTRM	scale for pitch rate demand from trim switch
SMX(4)	coefficients for the thrust to airspeed model
STICK	desk stick demand
STICKS	scaled stick demand (right-hand inceptor demand)
STKAUT	stick authority scaling factor
STKCNT	set to 1 if the stick is centered
SWITCH	switching point marker
TH2DEMZ	thrust trim demand from pitch attitude
THEJDX	control law nozzle servo demand
THETD	pitch attitude (degrees)
THRST(4)	states for the engine compensator
THRSTA	demand to the throttle control law
THRSTO	output from throttle dynamic compensator
TRMAUT	trim-switch command authority (degrees/s)
UB	horizontal body axis speed (feet/sec)
UOBSER(3)	plant inputs multiplied be the inverse of W1
UOBSIM(3)	equivalent plant inputs used to drive the observer
UDEMND(3)	output from W1
UPLANT(3)	achieved plant inputs used to drive Hanus form of W1
VBCLHI	top of blend between 3ip-3op and 2ip-2op control
VBCLLO	start of blend between 3ip-3op and 2ip-2op control
VBSTHI	top of stick blend from vertical 'g' to q demand
VBSTLO	start of stick blend from vertical 'g' to q demand
VENPHI	top of low engine limit blend (knots)
VENPLO	start of low engine limit blend (knots)
VGAUTH	stick authority for 'g' demands
VGN111	variable gain for X-loop
VGN222	variable gain for Z-loop
VGN333	variable gain for pitch-loop
VHDSWI	speed (knots) at which HUD switches between VSTOL and NAV modes
VHOR	horizontal ground speed (feet/sec)
VHORKT	horizontal ground speed (knots)
VKD	vertical speed (feet/sec)
VKFILT	filtered indicated airspeed (knots)
VKHUD	airspeed displayed onn the HUD
VKN	as VHOR
VKTAXF	speed signal derived from horizontal thrust (knots)

VKTIAS	indicated airspeed in knots
VTKT	airspeed (knots)
VTKTAP	derived airspeed signal used for feedback
WAP(4)	states for thrust to airspeed model
WB	vertical body axis speed (feet/sec)
WTONWH	weight on wheels flag
XAMDA1	temporary variable used when interpolating matrices
XAMDA2	temporary variable used when interpolating matrices
XKT2NG	scaling factor from knots to units of 'g'×s
XINSSW	flag set to 1 if INS ground speed signal available
XLAM	λ_x : airspeed to acceleration feedback ratio
XLHCOE(4)	coefficients for left-hand dynamics compensation
XLHRIN	variable used for implementation of LH dynamics
XLHRPU	previous output from LH dynamics compensator
XLHRS1	LH tuning for small demands
XLHRS2	LH tuning for large demands
XLHRSQ	left-hand precompensator non-linear lag
XM0(3,3)	align matrix for resolved thrusts
XMPDET	determinant of the align matrix
XNZDM1	unbounded nozzle angle demand
XNZDM2	unbounded rate limited nozzle angle demand
XPTDEM	scheduled pitch attitude demand for high transition
XTHLIM	authority limited XTHRTS
XTHROT	cockpit left-hand inceptor demand
XTHRTS	scaled pilot left-hand inceptor demand
XW1N(8)	state vector for the dynamic precompensator
XW1P(8)	previous values of XW1N(*)
YPLANT(3)	output of the shaped plant
ZEROZZ	datum for the incidence trim
ZLAM	λ_z : vertical speed to acceleration feedback ratio
α	angle of incidence (degrees)
γ	flight path angle (degrees), or $\mathcal{H}_\infty$ cost function according to context
θ	pitch attitude (degrees)
ϕ	bank angle (degrees)

APPENDIX B
BLOCK DIAGRAM

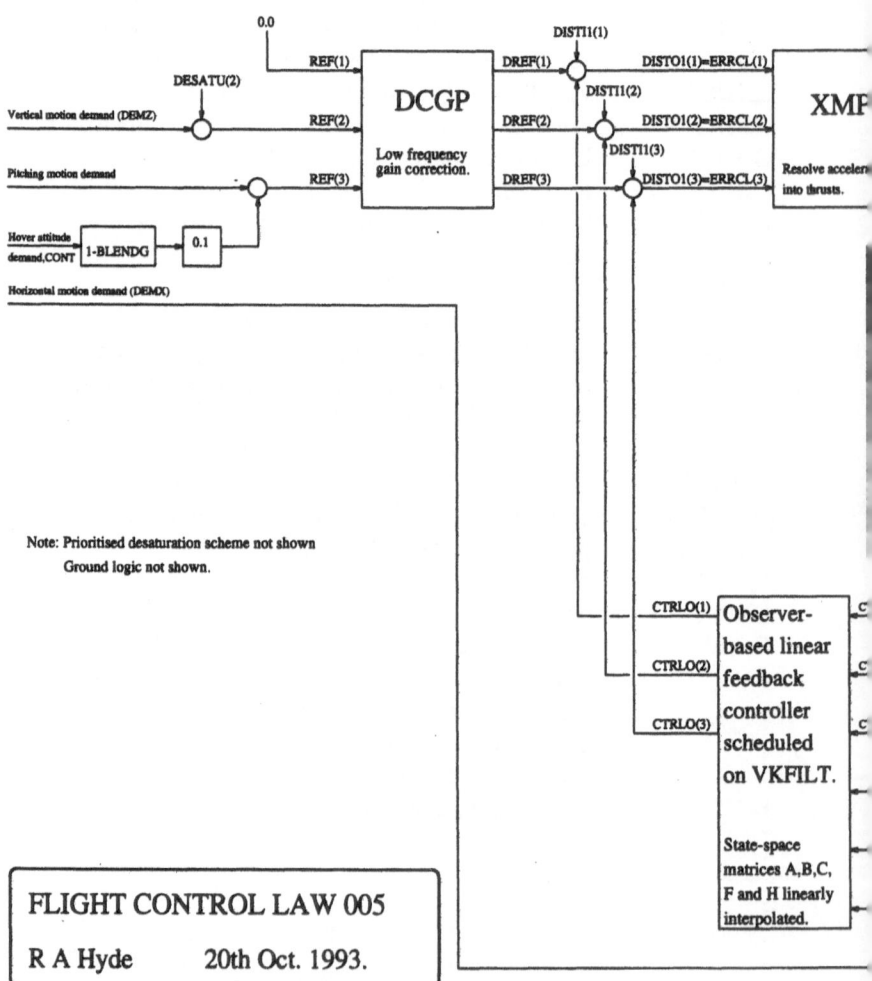

0.0

DESATU(2)

REF(1)

Vertical motion demand (DEMZ)

REF(2)

Pitching motion demand

REF(3)

Hover attitude
demand,CONT   1-BLENDG   0.1

Horizontal motion demand (DEMX)

DCGP

Low frequency
gain correction.

DREF(1)

DREF(2)

DREF(3)

DISTI1(1)

DISTI1(2)

DISTI1(3)

DISTO1(1)=ERRCL(1)

DISTO1(2)=ERRCL(2)

DISTO1(3)=ERRCL(3)

XMP

Resolve acceler
into thrusts.

Note: Prioritised desaturation scheme not shown
   Ground logic not shown.

CTRLO(1)

CTRLO(2)

CTRLO(3)

Observer-
based linear
feedback
controller
scheduled
on VKFILT.

State-space
matrices A,B,C,
F and H linearly
interpolated.

FLIGHT CONTROL LAW 005

R A Hyde   20th Oct. 1993.

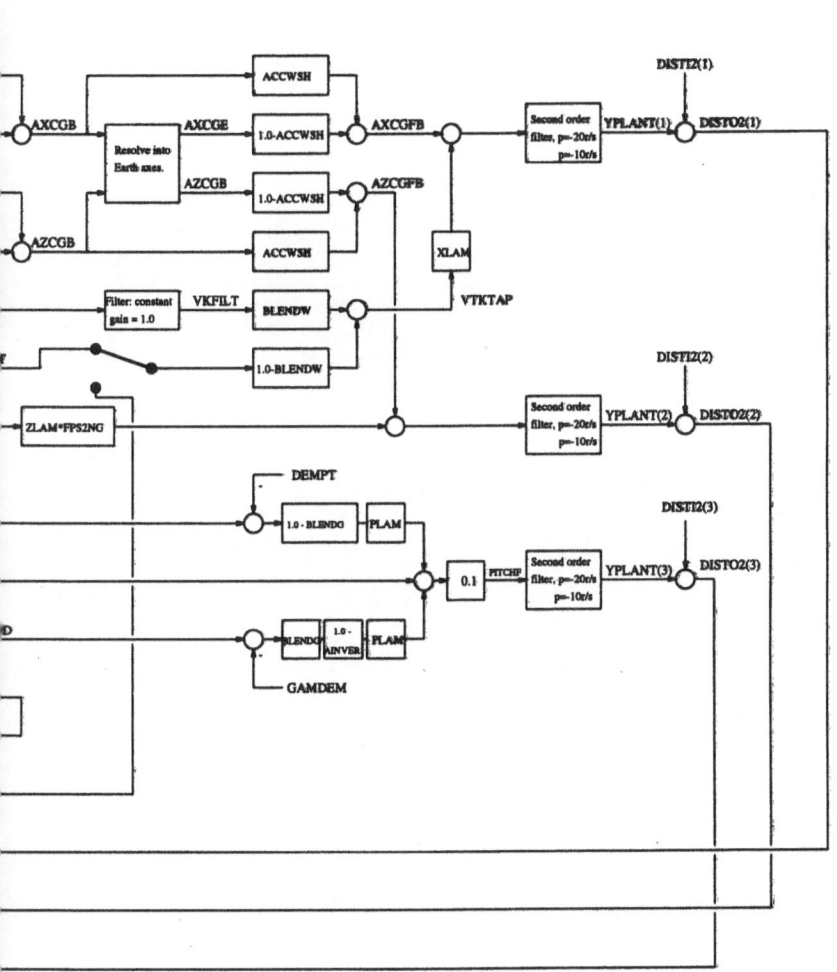

REFERENCES

1. G. Zames. Feedback and optimal sensitivity: Model reference transformations, multiplicative semi-norms, and approximate invereses. *IEEE Transactions on Automatic Control*, AC-23:301–320, 1981.
2. J. Doyle, K. Glover, P. Khargonekar, and B. Francis. State-space solutions to standard $\mathcal{H}_2$ and $\mathcal{H}_\infty$ control problems. *IEEE Transactions on Automatic Control*, 34(8):831–847, August 1989.
3. B.A. Francis. *A Course in $\mathcal{H}_\infty$ Control Theory*. Lecture Notes in Control and Information Sciences. Springer-Verlag, 1987.
4. R.A. Hyde. *The Application of Robust Control to VSTOL Aircraft*. PhD thesis, University of Cambridge, August 1991.
5. J.S. Freudenberg. Plant directionality, coupling and multivariable loop-shaping. *International Journal on Control*, 51(2):365–390, 1990.
6. R. Hanus, M. Kinnaert, and J.L. Henrotte. Conditioning technique, a general anti-windup and bumpless transfer method. *Automatica*, 23(6):729–739, 1987.
7. H.W. Bode. *Network Analysis and Feedback Amplifier Design*. New York: Van Nostrand, 1945.
8. A.G.J. MacFarlane and I. Postlethwaite. The generalised Nyquist stability criterion and multivariable root loci. *Internataional Journal of Control*, 25(1):81–127, 1977.
9. J.M. Maciejowski. *Multivariable Feedback Design*. Addison-Wesley, 1989.
10. S.J. Williams and P.R. Smith. A comparison of characteristic locus and $\mathcal{H}_\infty$ methods in VSTOL flight control system design. In *Proceedings of AIAA Conference on Guidance Navigation and Control*, pages 565–575, 1989.
11. G. Zames. On the input-output stability of time-varying nonlinear feedback systems–part I. *IEEE Transactions on Automatic Control*, AC-11(3):228–238, July 1966.
12. G. Zames. On the input-output stability of time-varying nonlinear feedback systems–part II. *IEEE Transactions on Automatic Control*, AC-11(3):465–476, July 1966.
13. G.J. Balas, J.C. Doyle, K. Glover, A. Packard, and R. Smith. *Mu-Analysis and Synthesis Toolbox for Matlab*. The Mathworks Inc, April 1991.
14. S.J. Williams. $\mathcal{H}_\infty$ for the layman. *Proceedings of the Institute of Measurement and Control*, 24:18–21, February 1991.
15. K. Glover and J.C. Doyle. State-space formulae for all stabilzing controllers that satisfy an $\mathcal{H}_\infty$-norm bound and relations to risk sensitivity. *Systems and Control Letters*, 11:167–172, 1988.
16. J.A. Sefton and K. Glover. Pole/zero cancellations in the general $\mathcal{H}_\infty$ problem with reference to a two block design. *Systems and Control Letters*, 14:295–306, 1990.
17. I. Postlethwaite, M.-C. Tsai, and D.-W. Gu. Weighting function selection in $\mathcal{H}_\infty$ design. In *Proceedings of IFAC Congress on Automatic Control*, 1990.

18. S.J. Williams and R.A. Hyde. A comparison of different $\mathcal{H}_\infty$ design methods for VSTOL flight control system design. In *Proceedings of American Control Conference*, pages 2508–2513, San Diego, May 1990.

19. K. Glover and D.C. McFarlane. Robust stabilization of normalized coprime factor plant descriptions with $\mathcal{H}_\infty$-bounded uncertainty. *IEEE Transactions on Automatic Control*, 34(8):821–830, August 1989.

20. D.C. McFarlane and K. Glover. A loop shaping design procedure using $\mathcal{H}_\infty$ synthesis. *IEEE Transactions on Automatic Control*, 37(6):759–769, 1992.

21. D.C. McFarlane. *Robust Controller Design Using Normalized Coprime Factor Plant Descriptions*. PhD thesis, University of Cambridge, 1988.

22. J.C. Doyle. Lecture notes on advances in multivariable control, October 1984. ONR/Honeywell Workshop on Advances in Multivariable Control, Minneapolis,MN.

23. A. Packard and J. Doyle. The complex structured singular value. *Automatica*, 29:71–109, 1993.

24. M.K.H. Fan, A.L. Tits, and J.C. Doyle. Robustness in the presence of mixed parametric uncertainty and unmodelled dynamics. *IEEE Transactions on Automatic Control*, 36(1):25–38, January 1991.

25. J.C. Doyle, K. Lenz, and A. Packard. Design examples using μ-synthesis: Space shuttle lateral axis FCS during reentry, 1987. NATO ASI Series, Volume F34.

26. G.J. Balas. *Robust Control of Flexible Structures: Theory and Experiments*. PhD thesis, California Institute of Technology, Pasadena USA, 1990.

27. J.A. Franklin. Control of V/STOL aircraft. *Aeronautical Journal*, pages 157–173, May 1986.

28. J.M. Ramsden. Towards Harrier III. *Aerospace*, pages 8–14, February 1991. Published by The Royal Aeronautical Society.

29. M. Gainza. Flying the Sea Harrier. *Pilot*, pages 29–34, February 1990. Pilot Publishing Company.

30. E.A.M. Muir and M.G. Kellett. The RAE generic VSTOL aircraft model — GVAM87. Technical report, Royal Aerospace Establishment, Bedford, 1987.

31. A.W. Babister. *Aircraft Dynamic Stability and Response*. Pergamon Press, 1980.

32. D. McLean. *Automatic Flight Control Systems*. Prentice Hall, 1990.

33. Anon. Flying qualities of piloted VSTOL aircraft. U.S. Military Specification MIL-F-833000, 1970.

34. Anon. Flying qualities of piloted airplanes. U.S. Military Specification MIL-F-8785C, 1980.

35. Anon. VSTOL handling - I - criteria and discussion. Advisory Group for Aerospace Research and Development R-577-70, North Atlantic Treaty Organisation, 1970.

36. J.C. Gibson. Evaluation of alternate handling qualities criteria in highly augmented unstable aircraft. *American Institute of Aeronautics and Astronautics*, 2844, 1990.

37. J.C. Gibson. Handling qualities for unstable combat aircraft. In *Proceedings of 15th Congress of the International Council of the Aeronautical Sciences*, volume 1, pages 433–445, London, September 1986.

38. B. Kouvaritakis. *Characteristic Locus Methods for Multivariable Feedback Systems Design*. PhD thesis, University of Manchester, 1974.

39. J.M. Ford, J.M. Maciejowski, and J.M. Boyle. *Matlab Multivariable Frequency Domain Toolbox*. The Mathworks Inc, July 1993. Version 2.4.

40. G. Vinnicombe. *Measuring Robustness of Feedback Systems*. PhD thesis, University of Cambridge, December 1992.

41. S. Skogestad, M. Morari, and J.C. Doyle. Robust control of ill-conditioned plants: High-purity distillation. *IEEE Transactions on Automatic Control*, 33(12):1092–1105, 1988.

42. J.C. Doyle and G. Stein. Multivariable feedback design: Concepts for a classical/modern synthesis. *IEEE Transactions on Automatic Control*, AC-26(1):4–16, February 1981.

43. R.Y. Chiang, M.G. Safonov, and J.A. Tekawy. $\mathcal{H}_\infty$ flight control system design with large parametric robustness. In *Proceedings of American Control Conference*, pages 2496–2501, San Diego, May 1990.

44. J.S. Freudenberg. Analysis and design for ill-conditoned plants. part 1. lower bounds on the structured singular value. *International Journal on Control*, 49(3):851–871, 1989.

45. Y.S. Ebrahimi and E.E. Coleman. Design of localizer capture and track using classical control techniques. *IEEE Control Systems Magazine*, 10(4):5–12, 1990.

46. D. Mustafa. *Minimum Entropy $\mathcal{H}_\infty$ Control*. PhD thesis, Cambridge University, May 1989.

47. O. Yaniv and N. Barlev. Robust non-iterative synthesis of ill-conditioned plants. In *Proceedings of American Control Conference*, pages 3065–3066, San Diego, May 1990.

48. R.A. Hyde, K. Glover, and S.J. Williams. Scheduling by switching of $\mathcal{H}_\infty$ controllers for a VSTOL aircraft. In *Proceedings of Applications of Multivariable System Techniques*, pages 33–42, Bradford, UK, 1990.

49. P.J. Campo, M. Morari, and C.N. Nett. Multivariable anti-windup and bumpless transfer : A general theory. In *Proceedings of American Control Conference*, pages 1706–1711, Pittsburgh, 1989.

50. P.J. Campo. *Studies in Robust Control of Systems Subject to Constraints*. PhD thesis, California Institute of Technology, 1989.

51. D.J.F. Hopper. *Active Control of VSTOL Aircraft*. PhD thesis, University of Salford, April 1990.

52. P. Kapasouris. Gain-scheduled multivariable control for the GE-21 turbofan engine using the LQG and LQG/LTR methodologies. Master's thesis, Massachusetts Institute of Technology, 1984.

53. M.G. Kellet. Continuous scheduling of $\mathcal{H}_\infty$ controllers for a MS760 Paris aircraft. In *Proceedings of Institute of Measurement and Control Symposium on Robust Control System Design*, Cambridge, March 1991.

54. J.S. Shamma and M. Athans. Guaranteed properties for nonlinear gain scheduled control systems. In *Proceedings of the 27th Conference on Decision and Control*, pages 2202–2208, Austin,Texas, December 1988.

55. J.S. Shamma and M. Athans. Analysis of gain scheduled control for non-linear plants. *IEEE Transactions on Automatic Control*, 35(8):898–907, August 1990.

56. A.C.M. Ran and L. Rodman. On parameter dependence of solutions of algebraic Riccati equations. *Mathematics of Control, Signals, and Systems*, 1:269–284, 1988.

57. D.C. McFarlane and K. Glover. *Robust Controller Design Using Normalized Coprime Factor Plant Descriptions*. Lecture Notes in Control and Information Sciences. Springer-Verlag, 1990.

58. R.A. Hyde. Control of multivariable systems in the face of saturation limits—a study on VSTOL aircraft. Technical Report 11, Cambridge Control Ltd., July 1988.

59. P. Kapasouris, M. Athans, and G. Stein. Design of feedback control systems for stable plants with saturating actuators. In *Proceedings of the 27th Conference on Decision and Control*, pages 469–479, Austin,Texas, December 1988.

60. M.J. Pélegrin. A new complement for air and spacecraft : a man/a computer. In *Proceedings of IFAC Congress on Automatic Control*, Munich, 1987.
61. R.A. Hyde and H.-D. Joos. Application of the Multi Objective Programming System MOPS to a VSTOL aircraft controller design. Technical Report TR R57–91, DLR Oberpfaffenhofen, Institut für Dynamik der Flugsysteme, July 1991.
62. C.N. Nett and J.A. Polley. Integrated design/implementation of nonlinear digital controllers. In *Proceedings of American Control Conference*, pages 1671–1678, 1988.
63. R.A. Hyde and K. Glover. A comparison of different scheduling techniques for $\mathcal{H}_\infty$ controllers. *Transactions of the Institute of Measurement and Control*, 13(5), 1991. Originally presented at the IMC Symposium on Robust Control System Design, Cambridge, March 1991.
64. H.-D. Joos. Informationstechnische Behandlung des mehrzieligen optimierungsgestützten regulungstechnische Entwurfs. Technical Report TR R44–91, DLR Oberpfaffenhofen, Institut für Dynamik der Flugsysteme, February 1991.
65. H.-D. Joos. MOPS — Multi Objective Programming System software documentation 1.1. Technical Report TR R43–91, DLR Oberpfaffenhofen, Institut für Dynamik der Flugsysteme, February 1991.
66. R.A. Hyde. Implementation complexity reduction of $\mathcal{H}_\infty$-optimal controllers using model matching and parameter optimisation. Technical Report TR R71–92, DLR Oberpfaffenhofen, Institut für Dynamik der Flugsysteme, December 1991.
67. O.P. Nicholas and C.M. Stephens. The VAAC/VSTOL flight control research project. *Aerospace*, July 1989. Published by The Royal Aeronautical Society.
68. P.A. Iglesias. *Robust Adaptive Control for Discrete-Time Systems*. PhD thesis, University of Cambridge, 1991.
69. W. Love. Control law 005 code translation. Technical Report COA-EAS-085, Defence Research Agency, Bedford, May 1993.

INDEX